Science Plus

General Editor Jenny Jones

Published by Collins Educational
An imprint of HarperCollins*Publishers*
77–85 Fulham Palace Road,
London W6 8JB

© HarperCollinsPublishers Ltd 1996

First published 1996
Reprinted 1996

ISBN 0 00 322463 5

The Science Plus writing team
Elizabeth Forth
Jenny Jones
Bob McDuell
Shirley Parsons
Gareth Price
Pamela Singh
Linda Welds

Designed by Chi Leung

Edited by Dodi Beardshaw

Picture research by Caroline Thompson

Artwork by Barking Dog, Russell Birkett, Tom Cross, Jerry Fowler, Hardlines, Martin Shovel

Printed and bound by Scotprint, Musselburgh.

Acknowledgements

Every effort has been made to contact the holders of copyright material, but if any have been inadvertently overlooked, the publishers will be pleased to make the necessary arrangements at the first opportunity.

Photographs

The publishers would like to thank the following for permission to reproduce photographs:
(T = top, B = bottom, C = centre, L = left, R = right)
Allsport/C Cole 20L, M Hewitt 25L, B Stickland 26, D Rogers 37, P Rondeau 50, S Ward 66T; Neill Bruce 56C; Peter Roberts Collection/Neill Bruce 56T; Bruce Coleman Ltd/K Rushby 70; Ronald Grant Archive 12L, 68; Sally & Richard Greenhill 8, 9, 11; The Hulton Deutsch Collection 43; Universal (courtesy Kobal) 6R; Amblin/Universal (courtesy Kobal) 12R, MGM (courtesy Kobal) 60; Andrew Lambert 69C, 71B; NASA 74L; Gareth Price 10, 27, 34C&TR, 35; Mick Hutson/Redferns 63, 71C; Rex Features Ltd 33, 40, 41, 65; Science Photo Library 6L, 7, 17, 18, 53R, 55, 69T, 71T, 73, 74R, 75; Shout Pictures 4, 5, 16, 44, 46, 49, 53C, 61, 67; FSP/R Benali/Liaison 20R, FSP/G Saussier 51; Tony Stone Images 19, 24, 25R, 28, 29, 36, 39, 48, 52, 66B; C & S Thompson 30, 31, 32, 34L&CR; Zefa Pictures Ltd 38, 54, 62, 64.
Cover Photograph:NASA

Contents

1 Dead or alive?

1.1 Rescue!

Mountain rescue teams go out in all weathers to bring people back from snow-covered mountain tops. The teams are made up of volunteers who risk their lives to save others.

Some of the people in the photo are already dead. How do the rescuers know if someone is still alive?

- *How could the rescue teams tell if the people are alive?*
- *Have you ever been hill walking? Where?*
- *Would you be willing to join a mountain rescue team?*
- *Do you think they should be paid for their work?*

When people are alive they **breathe**, their blood **circulates**, they grow, they move, they eat, **digest** and **excrete** food and they may **reproduce**. The things living things do are called **life processes**.

Name: Chris Davies
Consultant: Mr. Evans

Responses:
Good, eyes react to light; feet and fingers react to pain.

Urine:
500mls collected 10.34am; bladder full

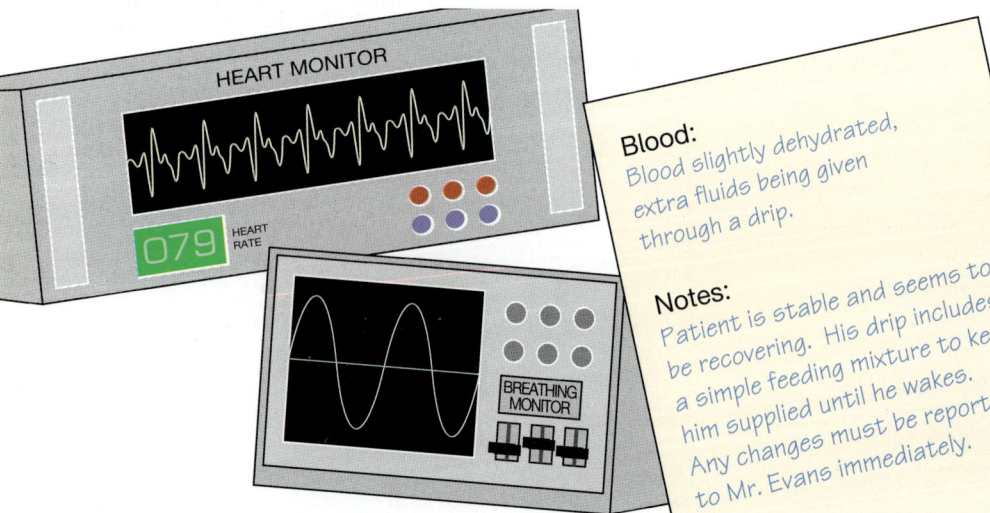

Blood:
Blood slightly dehydrated, extra fluids being given through a drip.

Notes:
Patient is stable and seems to be recovering. His drip includes a simple feeding mixture to keep him supplied until he wakes. Any changes must be reported to Mr. Evans immediately.

1 List the signs that show someone is alive.
2 What does the information on this chart tell you about this person's condition?
3 List all the processes that go on in a living body.
4 How can you tell that each process is working?

1.2 Kept alive

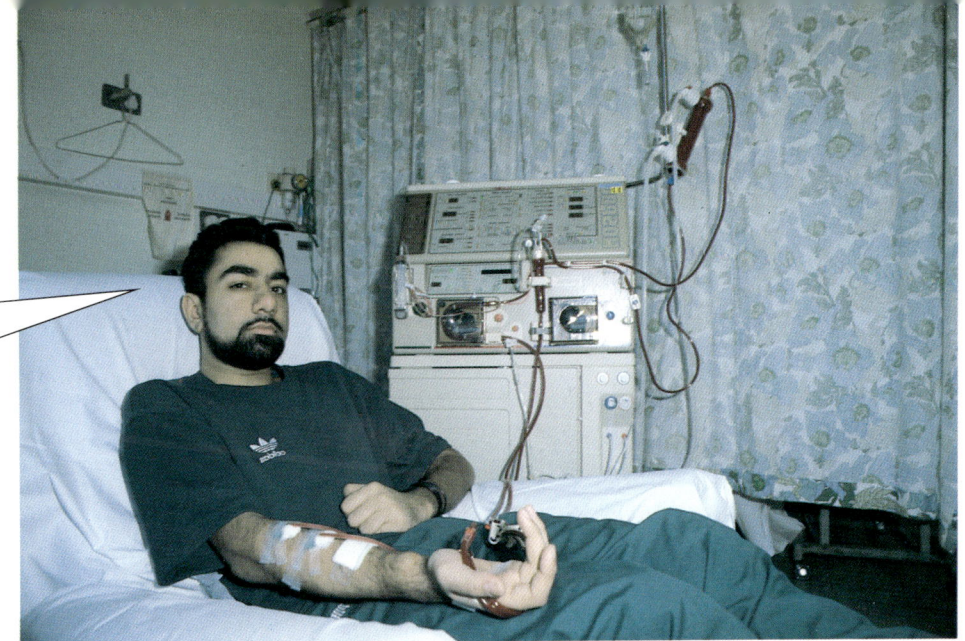

My kidneys were damaged by an infection about seven years ago. I am hoping to have a **transplant** operation soon. Meanwhile I have to use a kidney machine every week to clean wastes from my blood.

- *Would you give your organs for transplants? Why?*
- *Which organs would you donate? Some people would donate their kidneys but not their eyes.*

After we die, organs in our body can be transplanted into another person. The person who gives the organs is called the **donor**. The person who gets the organs is called the **host**.

When we die our bodies start to **decay**. After a short while the body is so badly damaged it cannot be used for transplants. Doctors often keep the body cool to reduce the speed of decay. Donors also need to be matched to hosts. This stops the host's body rejecting the new organ.

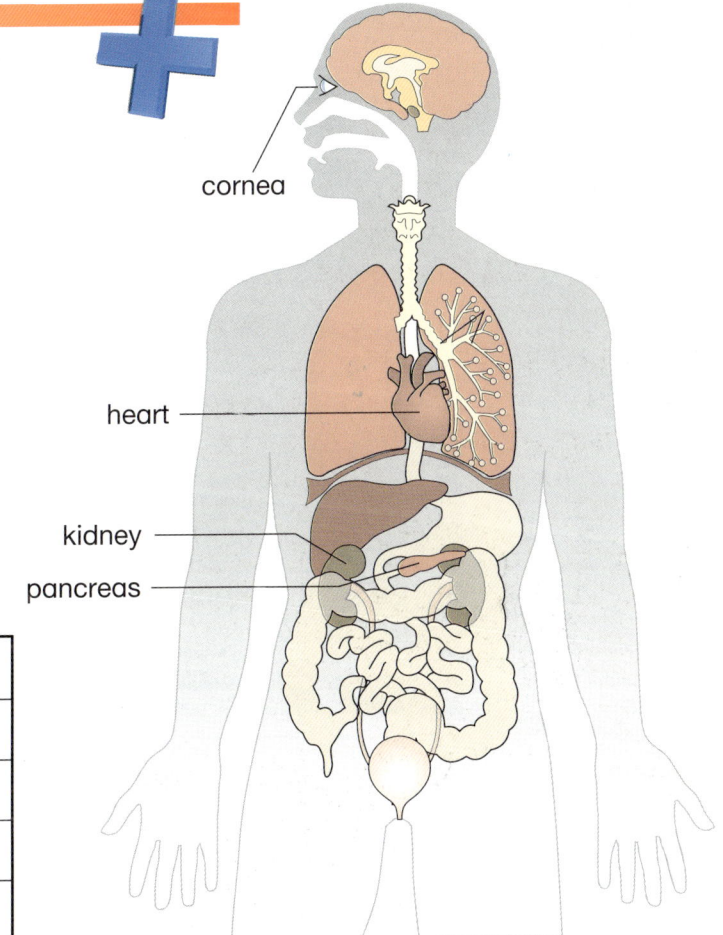

Donor organ	Cannot be used after:
heart	7 hours
liver	7 hours
pancreas	30 hours
kidney	72 hours

1 List six body parts that can be transplanted.
2 How are the removed organs kept in good condition?
3 What does the word donor mean?
4 What job does the kidney do in the body?
5 Which organ lasts the longest after death?
6 Design an organ donor card or a kidney carrying box.

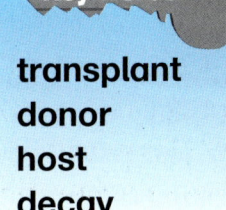

Key words

transplant
donor
host
decay

1.3 Artificial life

No-one has ever built an **artificial** human body. We cannot even make a robot that is alive in the way that we are. Scientists have made artificial limbs and are trying to make artificial hearts. In the future more body parts may be built.

- *Do you think it is a good idea to try to build artificial body parts? Why?*
- *Would you have an artificial heart if yours stopping working properly?*
- *If we can make artificial bodies we might be able to live forever. Do you think this would be a good thing?*

A robot does not have the same feelings and emotions as humans. It may not be as **sensitive**, but it must react to its environment. If it had artificial skin and got too close to a flame it might melt!

A **stimulus** is something which makes people react. What we do is called a **reaction**. So, when we touch something hot we pull our hand away quickly. Humans have very fast reactions to some stimuli. We call these reactions **reflexes**. Reflexes protect us from harm. For example, if dust gets up your nose, you sneeze. This stops the dust going down into your lungs. We don't have to think about what to do. We just react.

1 List the organs a humanlike robot would need.
2 What functions will these organs have to perform?
3 What is a reflex reaction?
4 List some reflexes the robot would need to survive.
5 Plan an investigation to test your reaction speed.

Key words

artificial
sensitive
stimulus
reaction
reflex

1.4 Alive forever?

Death and decay happen. We cool food and body organs to slow down the speed of decay. We cannot do this to live humans! We have to wait until they are already dead. Some people have their head frozen when they die. They hope that one day they can have a new body built back onto it!

Frozen assets

David and Paula Michael will not be buried or cremated when they die – they will be frozen. The Gateshead couple will be stored in liquid nitrogen until scientists work out how to revive them. In cold store their bodies will not decay for hundreds of years.

- *Would you like to be frozen after you die?*
- *Why are the bodies packed in ice and kept at -196°C?*

Bodies are built of **cells**. There are probably over a million, million cells in each of us. There are many different types of cell in the body. These cells look very different but they all have a **nucleus**, **cytoplasm** and a **cell membrane**.

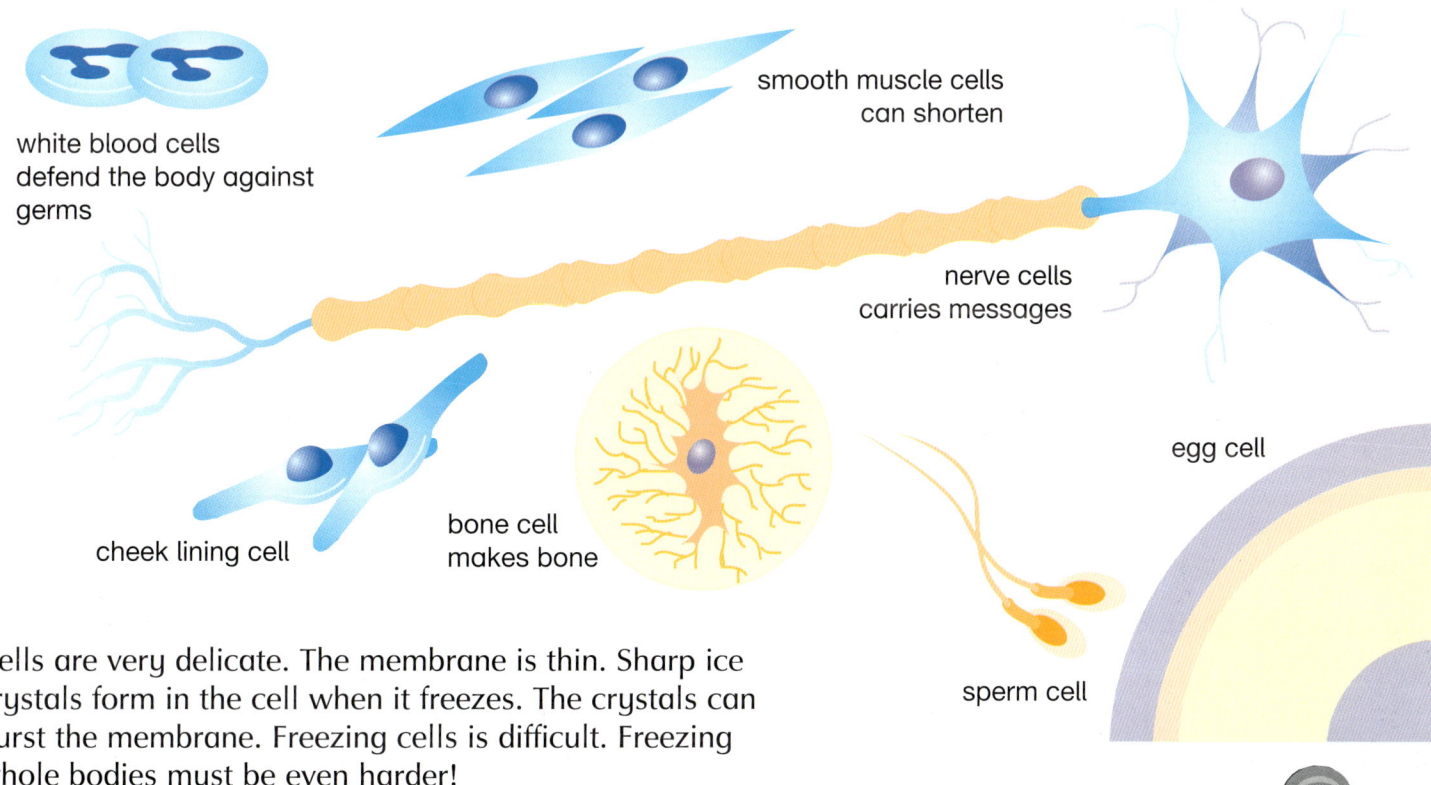

white blood cells defend the body against germs

smooth muscle cells can shorten

nerve cells carries messages

cheek lining cell

bone cell makes bone

sperm cell

egg cell

Cells are very delicate. The membrane is thin. Sharp ice crystals form in the cell when it freezes. The crystals can burst the membrane. Freezing cells is difficult. Freezing whole bodies must be even harder!

1 What is a cell?
2 Draw and label a cell.
3 List three types of human cell and say what they do.
4 What are some of the problems with trying to freeze cells?
5 Draw diagrams to show what happens when a cell freezes.

2 Babies

2.1 Making babies

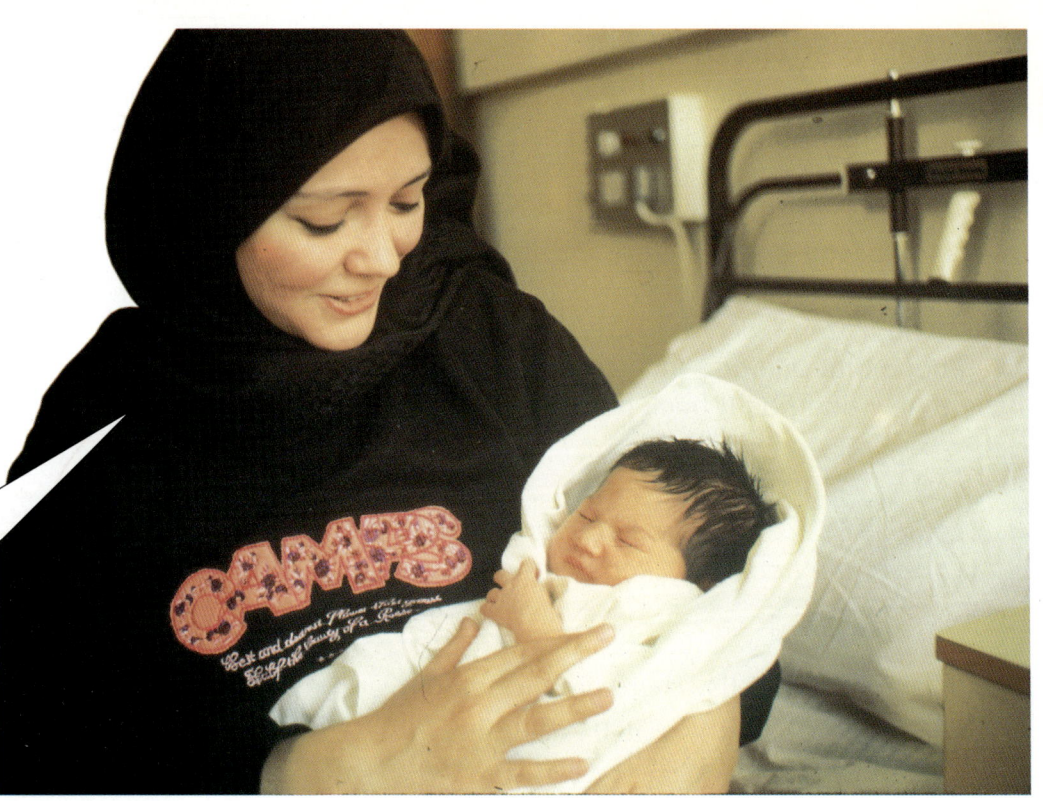

Having kids is a big responsibility. Friends who have kids told me about sleepless nights and dirty nappies. But they also spend all their time talking about their kids – and they often have more than one! This is my first child – isn't she beautiful?

- *Do you plan to have children?*
- *When is a good age to have children? Why?*
- *Why do you think people want kids?*

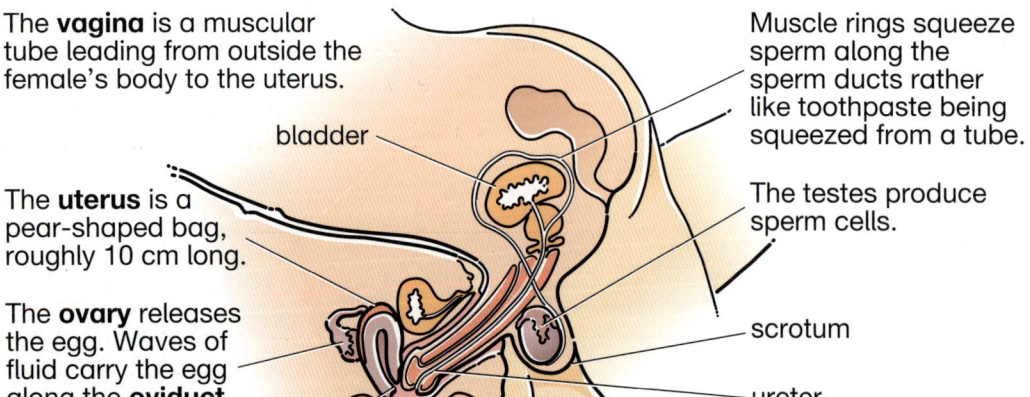

The **vagina** is a muscular tube leading from outside the female's body to the uterus.

bladder

The **uterus** is a pear-shaped bag, roughly 10 cm long.

The **ovary** releases the egg. Waves of fluid carry the egg along the **oviduct** towards the uterus.

cervix

Muscle rings squeeze sperm along the sperm ducts rather like toothpaste being squeezed from a tube.

The testes produce sperm cells.

scrotum

ureter

Once in the female, the **sperm** swim towards the oviducts. The sperm **fertilises** the egg in the oviduct.

1 Which part of the man produces sperm?
2 Which part of the woman produces the egg?
3 Where do the sperm and egg meet?
4 What moves the egg down the oviduct?
5 What moves the sperm along the sperm duct?

Key words

vagina
uterus
ovary
oviduct
sperm
fertilises

2.2 Before birth

People come to me at the antenatal clinic when they know they are **pregnant**. I measure the mum's **blood pressure**, **weight**, and **height**. The mother supplies the growing baby with all it needs through the placenta. I test the mother's urine for chemicals which show if the placenta is not working properly. The best result is always 'NAD' which means 'nothing abnormal detected'!

- *What sorts of questions would you ask if you visited an **antenatal** clinic?*
- *Do you think fathers should go as well as mothers? Why?*

	PREFERRED HOSPITAL............................		ANTENATAL RECORD			
	PREFERRED CONSULTANT....................					

DATE OF CONFINEMENT......*8/6/84*
PREGNANCY MATURITY......*40⁺⁶*......WEEKS
LABOUR 1st stage......*12hrs 45mins*
  2nd stage.........*1hr 40mins*
  3rd stage.........*5mins*
TOTAL DURATION......*14hrs 30mins*
METHOD OF DELIVERY
  Placenta and membranes complete
  Blood loss 300mls.
SEX: (MALE) FEMALE (ring as appropriate)
BIRTH MASS......*3350*......g
DISCHARGE MASS......*3600*......g
METHOD OF FEEDING......*Breast*
POST NATAL CARE HOSPITAL / (GP)

Date	Last period (weeks)	Can you hear baby's heart?	Urine tests		Blood pressure	Mass (kg)
			Glucose	Protein		
27/10/83	9			NAD	125/70	
16 Nov	11⁺⁶			NAD	100/60	48.2
8/12/83	(14⁺⁴)	FMF	NAD	NAD	115/60	50kg
6.1.84	19 wks	FMF	NAD	NAD	110/60	52kg
1 Jan 84	19⁺⁵		NAD	NAD	130/70	55.3
2.2.84	23	H	NAN	NAN	115/60	54 K
Mar 84	27⁺	FMF	NIL			
3/3/84	28⁴	FMF				
22.3				NAD		

1 What does an antenatal clinic do?
2 How long does a baby need to stay inside its mother before it is ready to be born?
3 What is the mother's weight on the 6th of January?
4 What does the **placenta** do?

Key words

pregnant
blood pressure
weight
height
antenatal
placenta

2.3 Birth

 They said it would start during the night – and they were right! I had gone to bed and was trying to sleep when the first **contraction** came. It didn't really hurt but it was enough to keep me awake. The contraction was the muscles tensing, getting ready to push the baby out. I knew that it was only the first sign of labour – a lot more had to happen first.

- *What things would be useful to take into hospital?*
- *Do you think the father should be there when the baby is born? Why?*
- *What sorts of things do you think the father can do during labour?*

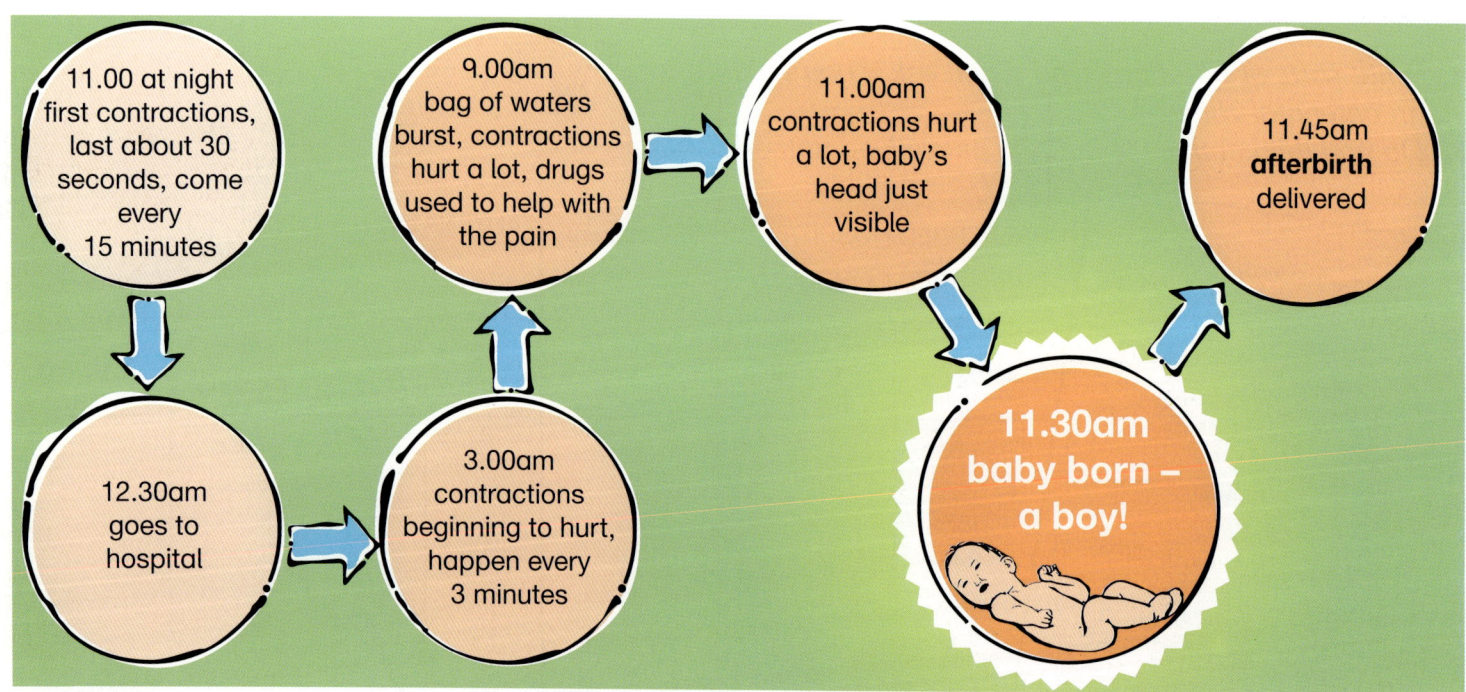

11.00 at night first contractions, last about 30 seconds, come every 15 minutes

12.30am goes to hospital

3.00am contractions beginning to hurt, happen every 3 minutes

9.00am bag of waters burst, contractions hurt a lot, drugs used to help with the pain

11.00am contractions hurt a lot, baby's head just visible

11.30am baby born – a boy!

11.45am **afterbirth** delivered

1   How long does it take from the start of labour to the birth?
2   How long after labour starts did this mother need drugs for the pain?
3   What is a contraction?
4   What do contractions do?

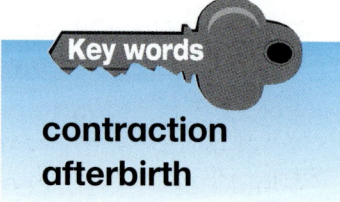

Key words

contraction
afterbirth

2.4 Health clinic

I am a community nurse. I visit the mum and her new baby every day for a week or so after they have come out of hospital. I keep an eye on mum and baby. I also make sure that mum knows about the local clinic. At the local clinic we weigh the baby. When the baby is six weeks old we do tests. This is to pick up any problems as early as possible. Later on we will check **eyesight** and **hearing**. One of the biggest advantages of the local clinic is that new mums can meet others in the same position. As you can see – it is very busy here!

- *Why do you think it is good for all new mothers and babies to visit their local clinic regularly?*
- *At the child health clinic the mother can also ask the health visitor or clinic doctor questions. What questions do you think a new mother might ask?*
- *Why is a record made of the baby's weight?*

WEIGHT RECORD (birth to 2 yrs)

Date	Weight	Weight gain	Comments
9/1/96	3.82kg	birth weight	
20/2/96	4.50kg	0.68kg	
19.3.96	4.95kg	0.45kg	reflexes ok
16/4/96	5.12kg	0.17kg	small weight gain
14/5/96	5.54kg	0.42kg	
11/6/96	5.95kg		hearing, sight ok

1 Use the information in baby Joe's card to plot a graph of his weight gain.
2 Why is there no figure for weight gain for the 9th of January?
3 What is the weight gain for the 11th of June?
4 Why do you think the health visitor has made a note on the record for the 16th April?

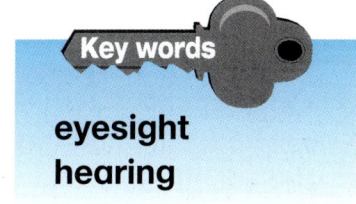

Key words

eyesight
hearing

3 Dinosaurs

3.1 Dinosaur fossils

- *Dinosaurs* are very popular in films and children's stories. Why do you think this is?
- Have you ever been to a museum where there are *fossils*? What did you see?
- Did what you saw in the museum help you to imagine what life was like then?

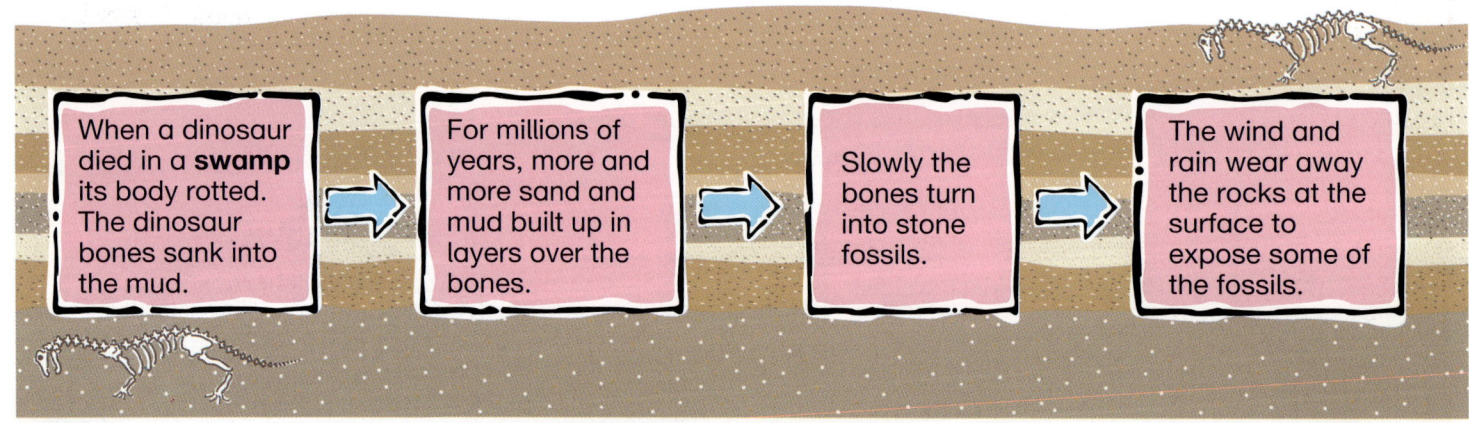

When a dinosaur died in a **swamp** its body rotted. The dinosaur bones sank into the mud.

→

For millions of years, more and more sand and mud built up in layers over the bones.

→

Slowly the bones turn into stone fossils.

→

The wind and rain wear away the rocks at the surface to expose some of the fossils.

The real Jurassic Park was in Texas 200 million years ago. Many different types of dinosaur lived there. We know this because of the fossils found in the rocks. The oldest rocks contain the oldest fossils. Usually, older rocks are underneath younger ones. We can tell the age of a rock, and the fossil it contains, by how deeply it is buried and the radioactive chemicals it contains.

1 List the things that happen when a fossil forms.
2 Draw a cliff showing three layers of rock.
3 On your cliff diagram, label the oldest layer.
4 On your cliff diagram, label the layer that will have the youngest fossils.
5 There are three ways to decide the age of a rock. What are they?

Key words

dinosaur
fossil
swamp

3.2 Information from fossils

Scientists put fossil bones together like a jigsaw. The trouble is, many of the pieces are missing or damaged – and there may be pieces from different jigsaws mixed in together.

- *Do you think it would be harder or easier to do the jigsaw without the picture?*
- *What clues would a scientist need to piece together fossil bones?*

Apatosaurus

Tyrannosaurus

Dinosaurs were different from each other. **Tyrannosaurus** was a hunter. **Ichthyosaurus** swam like a fish. They were **adapted** to live in a different places and feed in different ways.

Even dinosaurs of the same type would be slightly different from each other. One Tyrannosaurus might be slightly taller, or stronger than all the others living in that area. This might mean he would be a better hunter and so get more food. He might be more likely to survive and breed to produce stronger, taller dinosaurs. Over millions of years these changes might lead to a completely new sort of dinosaur.

Ichthyosaurus

1 List the adaptations the Ichthyosaur had for life in water.
2 List the adaptations the **Apatosaurus** had for life on land.
3 List the adaptations that the Tyrannosaurus had for life as a hunter.
4 How do you think a longer neck would help an Apatosaurus survive?
5 How do you think being able to run faster would help a Tyrannosaurus survive?

Key words

Tyrannosaurus
Ichthyosaurus
adapt
Apatosaurus

3.3 Flying dinosaurs

Pterodactyls were able to fly. One finger on each **forelimb** grew longer to support a large flap of skin. These were their wings. Their bones were light and strong. Their nests were in high places so that they could launch themselves into the air. They used the **air currents** to fly like gliders. They must have looked like giant, leathery paper aeroplanes.

- *Have you ever made paper aeroplanes?*
- *What shape do you think would make the best plane?*

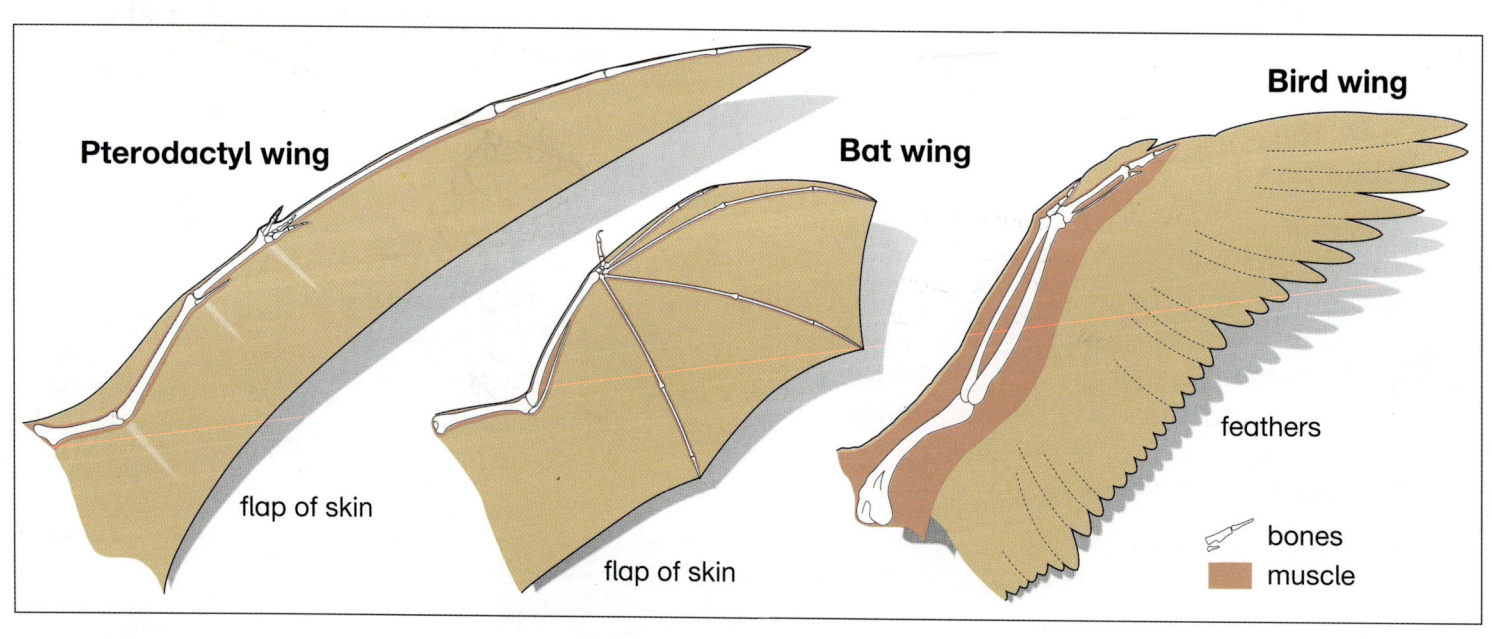

Pterodactyl wing

flap of skin

Bat wing

flap of skin

Bird wing

feathers

bones

muscle

1 How are pterodactyls like bats and birds?
2 Why do you think the bones of birds, bats and pterodactyls are light?
3 Plan an investigation to find the best shape for a pterodactyl wing.

Key words

pterodactyl
forelimb
air currents

3.4 Can Jurassic Park happen?

Your body **cells** contain **genes**. Genes control the way you grow and develop. If you have blue eyes it is because you have genes for blue eyes. Since genes control how living things grow, can we use dinosaur genes to grow a dinosaur? But where can we find dinosaur genes? There are no dinosaurs alive today. Dinosaurs are **extinct**.

- *Do you think scientists should try to recreate extinct dinosaurs? Explain your answer.*
- *What other animals from the past would you like to recreate?*
- *Would you like someone to grow another copy of you from your genes?*

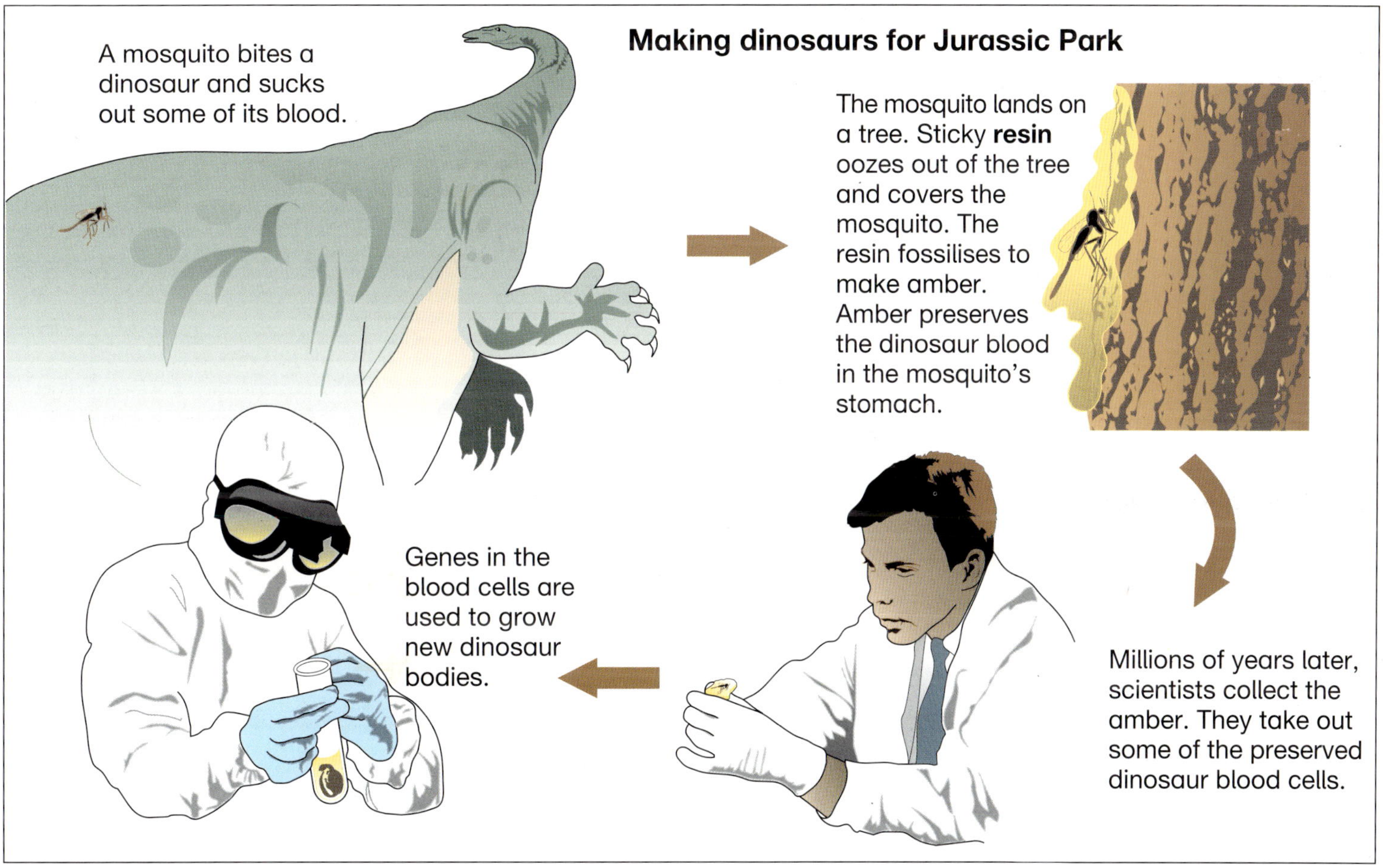

Making dinosaurs for Jurassic Park

A mosquito bites a dinosaur and sucks out some of its blood.

The mosquito lands on a tree. Sticky **resin** oozes out of the tree and covers the mosquito. The resin fossilises to make amber. Amber preserves the dinosaur blood in the mosquito's stomach.

Millions of years later, scientists collect the amber. They take out some of the preserved dinosaur blood cells.

Genes in the blood cells are used to grow new dinosaur bodies.

1. What does extinct mean? Make up your own sentence using the word extinct.
2. If a mosquito sucked human blood what genes would it have in its stomach?
3. List the reasons why people should try to recreate extinct dinosaurs.
4. List the reasons why people should not try to recreate extinct dinosaurs.

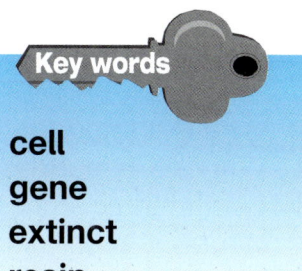

Key words

cell
gene
extinct
resin

4 Casualty

4.1 Accident!

- *Have you ever seen a serious accident? What happened? Could you help?*
- *How would you check to see if people are breathing?*

A person cannot live without **oxygen** for more than three minutes. When we stop **breathing** we stop taking in oxygen. The blood carries oxygen from the lungs to other parts of the body. If the heart stops beating then the blood cannot flow. If you lose a lot of blood or you stop breathing or your heart stops beating then the oxygen cannot get around the body. You could die.

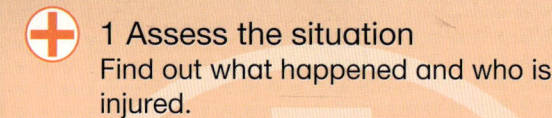

1 Assess the situation
Find out what happened and who is injured.

2 Make safe
Make sure that the person is no longer in danger and that no-one else is going to get hurt.

3 Give emergency first aid
a Airways – check that they are open. Remove a blockage if necessary.
b Breathing – check to see if the casualty is breathing. Give the kiss of life if necessary.
c **Circulation** – is the heart still beating? If yes, then check for signs of **bleeding**. Stop the flow of blood.

4 Get help
Get qualified help. If possible, someone should stay with the casualty.

1 Why is it important that the casualty keeps breathing?
2 Why is it important that the heart keeps beating?
3 List the three stages of emergency aid.
4 Why is it important to make sure the area is safe?
5 List the information you would need to give if you dialled 999.

Key words

oxygen
breathe
circulation
bleed

16

4.2 The liquid of life

This person's life has been saved by someone he has never met. The man needs blood to treat a disease called sickle cell anaemia. The red blood cells in sickle cell anaemia are the wrong shape and do not work properly. This man's life saver is a **blood donor**.

- *You have to be over 16 years of age to give blood. Do you think this is fair? Would you be willing to give blood?*
- *Have you ever needed a **blood transfusion**? Why?*

Red bone marrow makes **red blood cells**. Red blood cells pick up oxygen in the lungs and carry it around the body. These unusual cells do not have a nucleus and die after about 100 days.

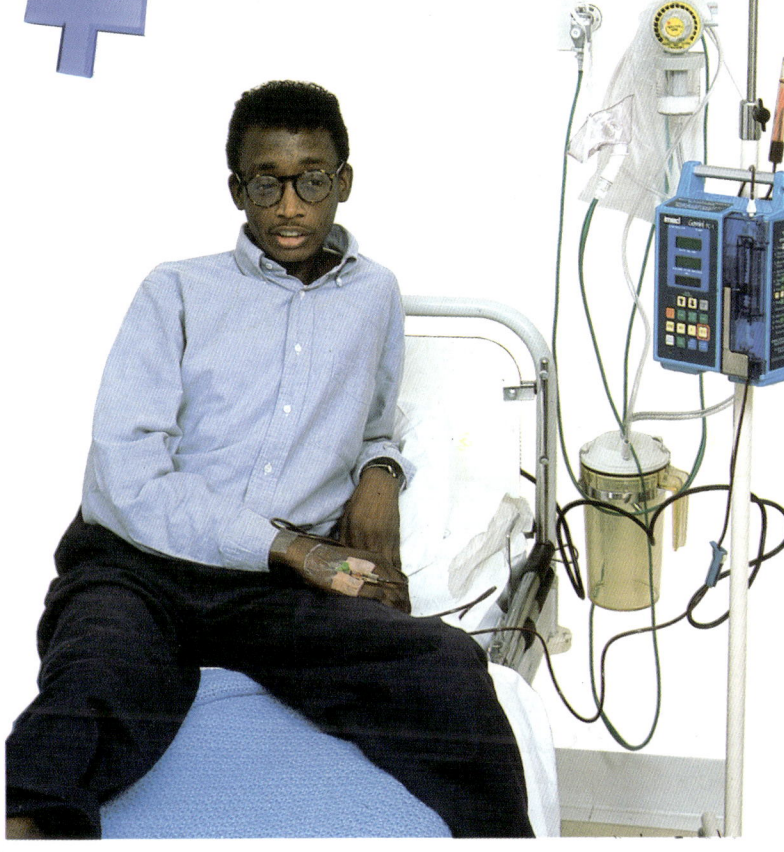

There are many different types of **white blood cells**. They all help to protect the body from disease.

Plasma is the liquid part of the blood. It is a pale yellow solution of sugar, salt and many other substances.

Platelets are very small particles in the blood. They help the blood to clot.

1 What do red blood cells do?
2 What do white blood cells do?
3 Which part of the blood helps it to clot?
4 List the things in plasma.
5 Carry out a survey to find out how many people have had a blood transfusion.
6 Design a poster that encourages people to become blood donors.

Key words

blood donor
blood transfusion
red blood cells
white blood cells
plasma
platelets

4.3 Heartbeat

Your heart pumps blood around the body. **Valves** in the heart control the way the blood flows. If you listen to the heart you can hear these valves opening and closing.

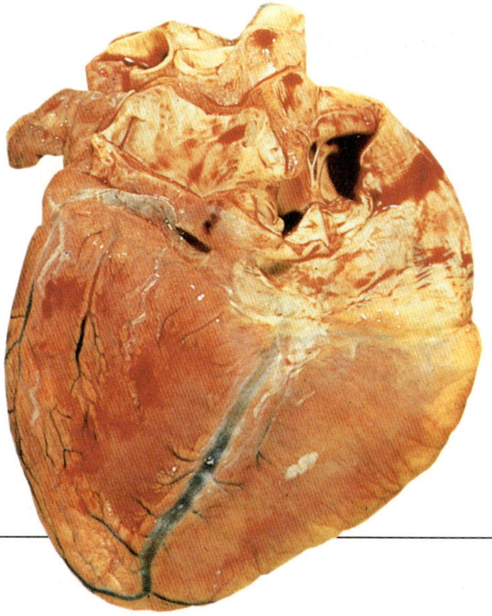

- *What is your pulse when you are sitting down?*
- *Where can you feel your pulse?*

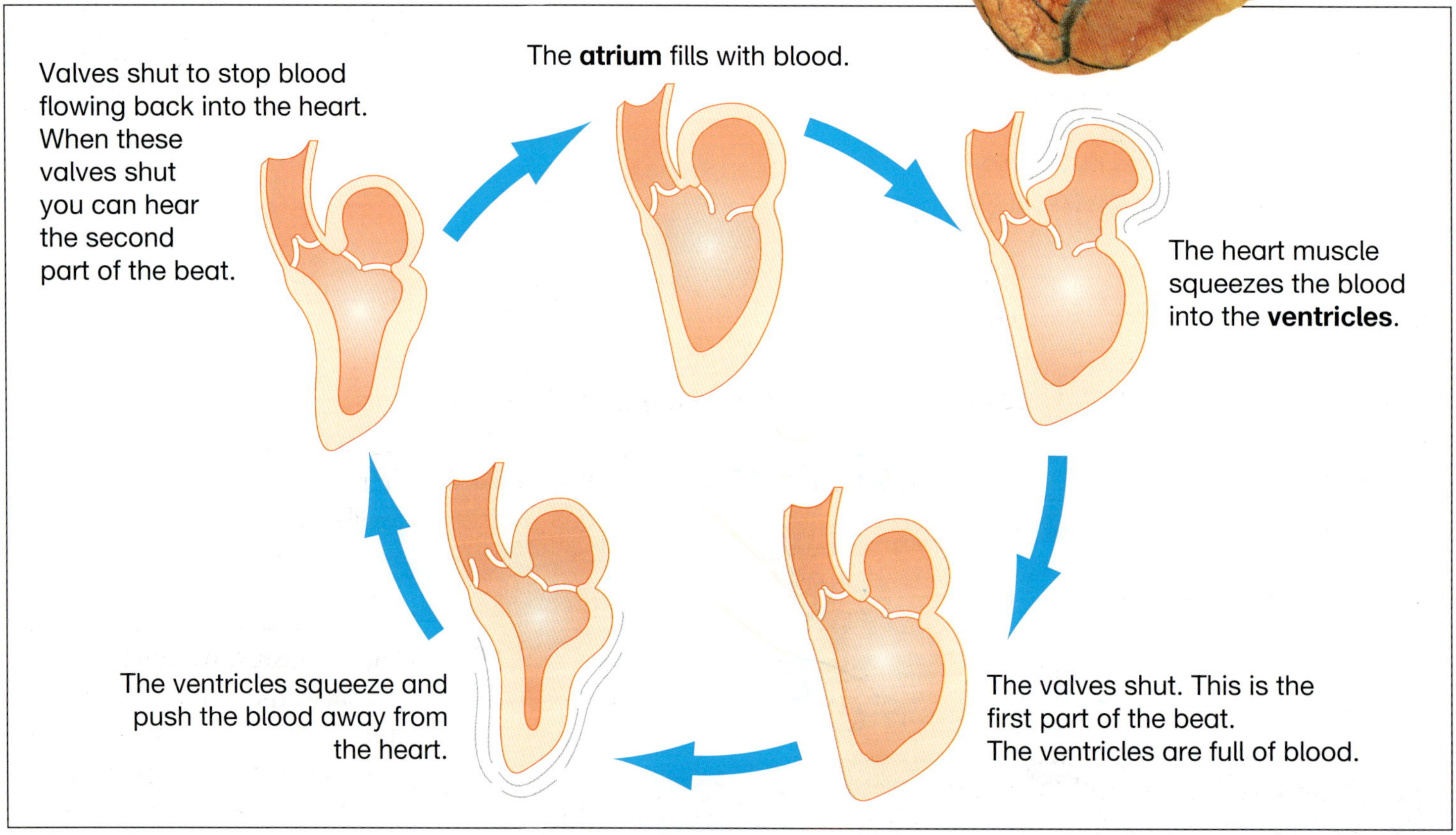

The **atrium** fills with blood.

Valves shut to stop blood flowing back into the heart. When these valves shut you can hear the second part of the beat.

The heart muscle squeezes the blood into the **ventricles**.

The ventricles squeeze and push the blood away from the heart.

The valves shut. This is the first part of the beat. The ventricles are full of blood.

The heart is really two pumps stuck together. The left side pumps blood from the body to the lungs. The right side takes blood from the lungs and pumps it to the body. In the lungs the blood loses **carbon dioxide** and gains **oxygen**. We breathe out the carbon dioxide. The blood carries the oxygen to other parts of the body.

1 Where does blood enter the heart?
2 How many sets of valves are there in each side of the heart?
3 What do the valves do?
4 What happens to the blood in the lungs?

Key words

valve
atrium
ventricle
carbon dioxide
oxygen

4.4 Lifelines

This old man is doing his morning exercises. He comes from China. Chinese people have very low rates of heart disease. They are much less likely to die from a **heart attack** than someone in Britain.

- *How much exercise do you get every week?*
- *Do you think this is enough?*
- *Do you think people should be forced to take enough exercise to keep them healthy? Why?*

The blood leaves the heart in tubes called **arteries**. The arteries divide into smaller and smaller tubes. The smallest tubes are called **capillaries**. These are so small you cannot see them without a microscope. The capillaries link up into larger and larger tubes to take blood back to the heart. These tubes are called **veins**.

As we get older our hearts begin to weaken.

Smoking damages the lungs. Chemicals in cigarette smoke also make the heart beat more quickly all the time.

Lack of **exercise** means muscles go flabby. Since the heart is a muscle it can get weaker if it becomes unfit. Exercise encourages the heart to grow and develop.

People who are overweight put a strain on their heart. The blood has to travel further around their body than in thinner people.

Sometimes a fatty substance coats the inside walls of blood vessels. This makes the **blood vessel** narrower. The heart has to work much harder to pump blood along it.

1 List the things that can damage your heart.
2 Sort your list into things we can prevent and those that we cannot do anything about.
3 List ways to keep your heart healthy.
4 Exercise makes your heart beat faster. Plan an investigation to find out how different sorts of exercise affect your heart rate.

Key words

heart attack
artery
capillary
vein
blood vessel
exercise

5 Building bodies

5.1 Special bodies

We are what we eat. We use our food to build us up and give us **energy**. The food we eat is called our **diet**. If there is something wrong with our diet there will be something wrong with our body. Of course, different people need different diets!

- *How do you think the Sumo wrestler's diet will differ from your diet?*
- *Do you think the wrestler or the body builder has a healthier body? Why?*

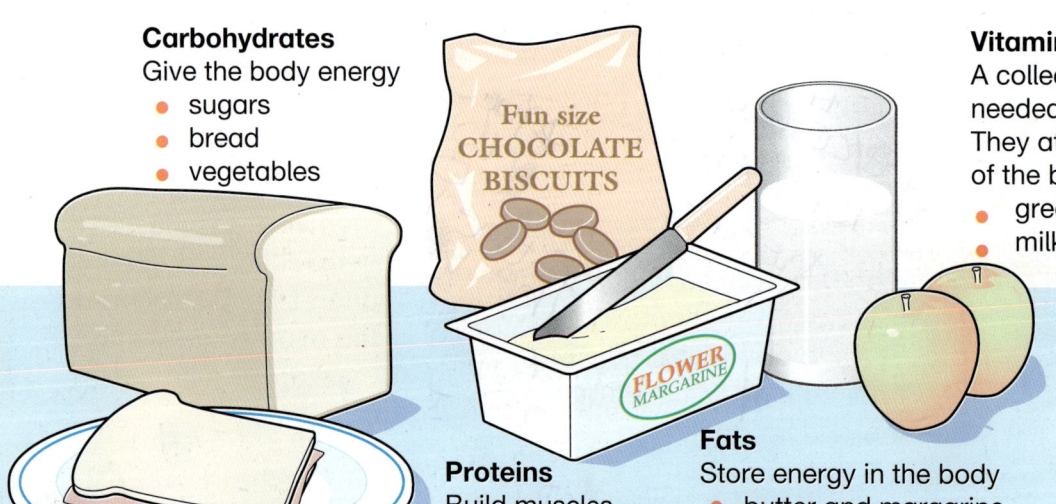

Carbohydrates
Give the body energy
- sugars
- bread
- vegetables

Fun size CHOCOLATE BISCUITS

FLOWER MARGARINE

Vitamins and minerals
A collection of different chemicals needed in very small amounts. They affect many different parts of the body.
- green vegetables and fruit
- milk and milk products

Proteins
Build muscles
- lean meat
- beans

Fats
Store energy in the body
- butter and margarine
- chocolate
- fatty meat

Key words

energy
diet
carbohydrates
vitamins
minerals
proteins
fats

1 List the foods you think the Sumo wrestler would eat.
2 List the foods you think the body builder would eat.
3 List the food you ate yesterday.
4 Sort your food into carbohydrates, fats and proteins.
5 Are you getting a good balance of the different food types?

5.2 Special foods

Most foods are a **mixture** of different types. Chicken has lots of **protein** and little **fat**. Beef has lots of protein but more fat. I need the protein but not the fat. So I eat more chicken. I buy special foods for body builders. I spend more time reading the labels than I do cooking. I take vitamin and mineral pills because I do not always eat enough fruit and vegetables. I work hard getting my body in shape.

- Are you careful about the foods you eat?
- Do you read the labels on food packs? Why?
- Do you eat foods that you know are bad for you? Or good for you?

4 STEWING STEAKS

Typical values per 100g (3.5oz)

ENERGY	223kcal, 932kJ
PROTEIN	13.9g
CARBOHYDRATE	0.0g
FATS	11.0g

Skinless Frozen Chicken

Typical values per 100g (3.5oz)

ENERGY	210kcal, 880kJ
PROTEIN	33.2g
CARBOHYDRATE	0.0g
FATS	5.4g

Chocolate Biscuit (KitKat)
Typical values per 100g
ENERGY 503kcal, 2103kJ
PROTEIN 7.6g
CARBOHYDRATE 59.1g
FATS 26.2g

INGREDIE
HYDROGE
WHEY PO
FLAVOU
ETABLE
THIS P

Roasted Salted Peanuts
Typical values per 100g
ENERGY 600kcal, 2489kJ
PROTEIN 29.0g
CARBOHYDRATE 8.6g
FATS 50.0g

1. Almost all foods are mixtures. Explain what the word mixture means.
2. Is it a good idea to eat a diet of only one type of food? Why?
3. Sort the foods shown here so that the one with the most protein comes first.
4. Draw a bar chart to show the carbohydrate, fat and protein in chicken.
5. Which foods would the body builder try not to eat? Why?

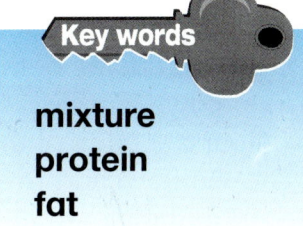

Key words

mixture
protein
fat

5.3 Big gut

Food must be digested before it is **absorbed**. **Digestion** means to break the food down into small particles. Digestion goes on in our **gut**. The gut is the tube that runs from the mouth to the **anus**. The body has to absorb food before it can be used.

- *Can you point out where your **stomach** is in your body? Try it.*
- *Where is the small **intestine**?*

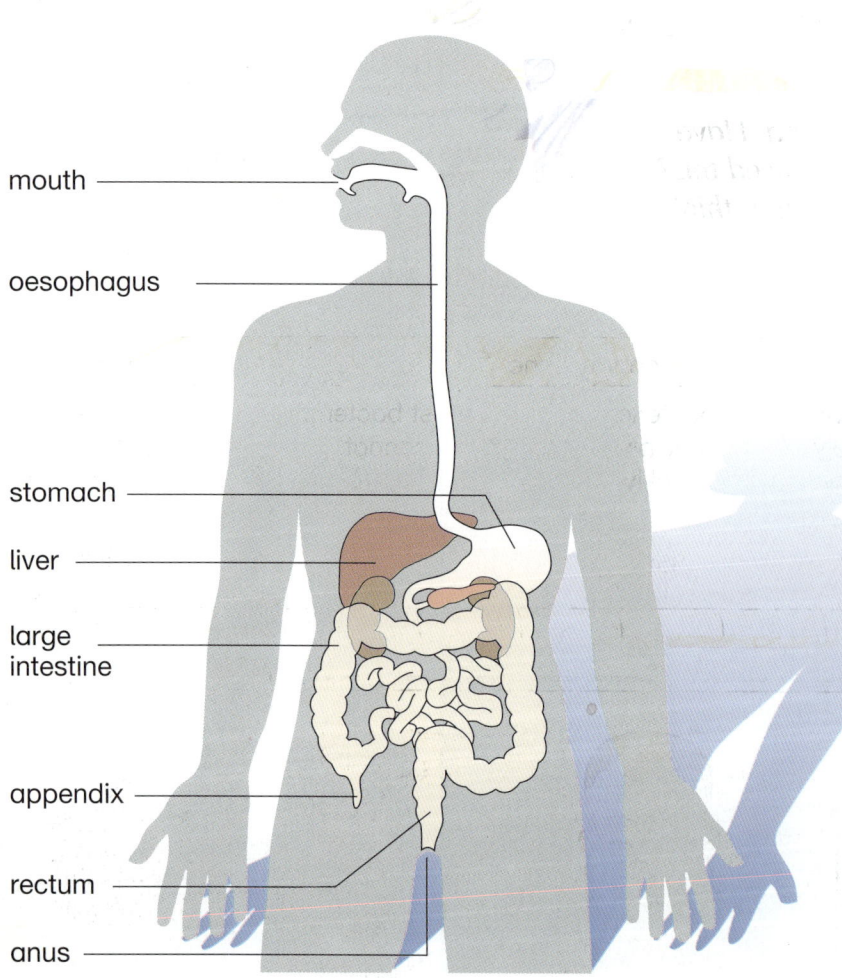

mouth

oesophagus

stomach

liver

large intestine

appendix

rectum

anus

The mouth breaks food into smaller lumps. Saliva helps to break down starchy foods to sugar

Acid and enzymes in the stomach break down the protein. The acid kills any bacteria in the food. Here the food is a watery sludge.

The small intestine breaks down foods. It can also absorb food into the blood.

The large intestine absorbs water from the food. The food begins to be more solid.

The rectum stores food. The ring of muscle in the anus holds wastes in the body until you go to the toilet.

1 List the parts of the gut that the food passes through from mouth to anus.
2 Draw an outline of the gut. Colour the parts that break food down in red.
3 Colour the parts of the gut that absorb food in blue.
4 Draw a flowchart to show what happens to food in the gut. Start in the mouth and finish when the food passes out of the body.

Key words

absorb
digestion
gut
anus
stomach
intestine

5.4 Gut rot

Eating or drinking too much, some **bacteria** or a **fever** can make you sick. These things can also give you the runs. Sickness and **diarrhoea** is the body's way of trying to get rid of the bad things, but it loses lots of water and salts too. Muscles that are working hard need more blood than usual. After a meal lots more blood goes to the gut to absorb the digested food. The muscles cannot get the blood they need! They hurt! We call this **cramp**.

Constipation happens when the undigested food stays in the large intestine for too long. It gets dried out and is hard to push out of the body. Eating more **fibre** in the diet may help.

- If we do not look after our gut it can go wrong. Have you ever been sick or had the runs? What caused this?
- Have you ever had cramp? When? What do you think caused it?

safe			danger zone		safe
bacteria cannot multiply	bacteria multiply very slowly	bacteria multiply slowly	bacteria multiply quickly	most bacteria cannot multiply	most bacteria die

-18°   0   16°   35°   100°

1. List the things that might have upset the body builder in the cartoon.

Look at the information given in the chart and answer the questions below.

2. What is the danger zone for bacteria?
3. Why is it dangerous to leave food out in a warm room?
4. How could the body builder kill the bacteria in his food?

6 Control systems

6.1 Controlling blood gases

Self-Contained Underwater Breathing Apparatus lets divers swim like fish. The SCUBA equipment uses an **aqualung** which contains pressurised air. The diver breathes air from this to stay underwater for hours.

- *How long can you hold your breath?*
- *Have you ever been diving?*
- *Would you like to? Why? Why not?*
- *What is the aqualung for?*

Your body needs energy to do anything. It gets energy from a chemical reaction called **respiration**. Respiration uses up sugar and **oxygen** and gives out **carbon dioxide** and water. Your lungs swap oxygen for carbon dioxide in the air you breathe.

Your body keeps the levels of oxygen and carbon dioxide within certain limits. When you run faster or exercise you need more energy. This means you need more oxygen. The body breathes more deeply to collect more oxygen from the air.

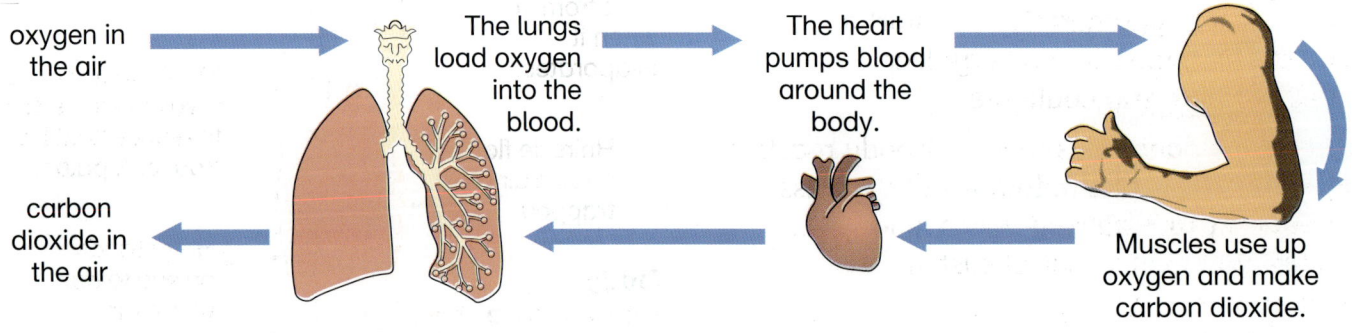

oxygen in the air → The lungs load oxygen into the blood. → The heart pumps blood around the body. → Muscles use up oxygen and make carbon dioxide.

carbon dioxide in the air ←

1　What is respiration?
2　Name two chemicals that respiration makes.
3　Why does the body need to carry out respiration?
4　How does the body get rid of the wastes from respiration?
5　A diver who swims very fast uses up the air in his aqualung sooner than one who swims more slowly. Why?

6.2 Controlling temperature

For some people getting too hot is the problem – for others it is getting too cold. To stay alive our body temperature must be controlled. Being too hot or too cold can kill.

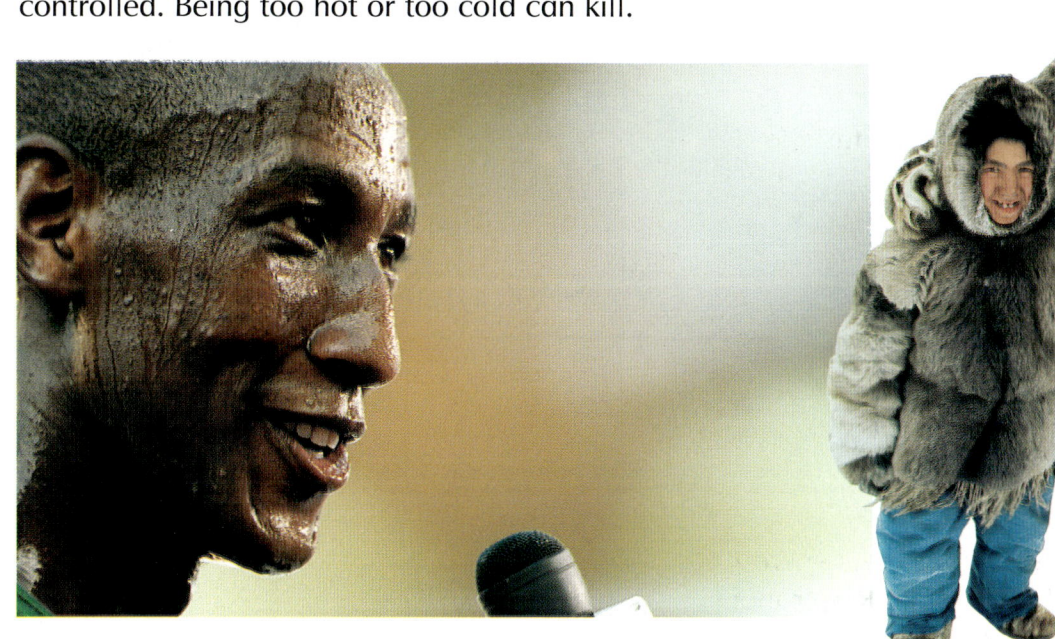

- *Have you had a holiday in a hot place? How did you keep cool?*
- *Have you been somewhere very cold? How did you keep warm?*
- *What happens to your body when you get cold?*

Your body works best at 37°C. If you get too cold you could get **hypothermia**. You start to feel sleepy and can pass into a very deep sleep called a **coma**. You could die.

Getting too hot is dangerous too. Your body reacts to changes in temperature automatically. Unless you are ill your **temperature** stays around 37°C. This is part of the body's control system – it balances your temperature.

Hot

Blood flows to the skin surface so it cools down in the air. You look redder.

Sweat is released. It takes heat from you when it evaporates.

Hairs lie flat so less air is trapped.

Body temperature

Cold

The blood flow to the ears and nose is reduced to cut down heat loss.

Muscles **shiver** to create heat.

Most blood keeps away from the skin to reduce heat loss. You look paler.

Hair stands on end to trap warm air.

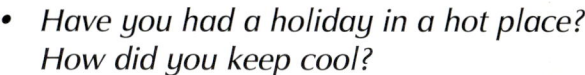

1 What is a healthy body temperature?
2 What is hypothermia?
3 List the things your body does to cool you down in hot weather.
4 List the things your body does to warm you up in cold weather.

Key words

hypothermia
coma
temperature
sweat
shiver

6.3 Controlling water levels

Marathons are hard work – and you lose pints of water during a summer run. If you want to win you've got to replace that water – but there's no time to stop running!

- *List the ways you take in water.*
- *What sorts of things make you **thirsty**?*
- *Do you drink more on a hot day than a cold day?*
- *How does your body get rid of extra **liquid**?*

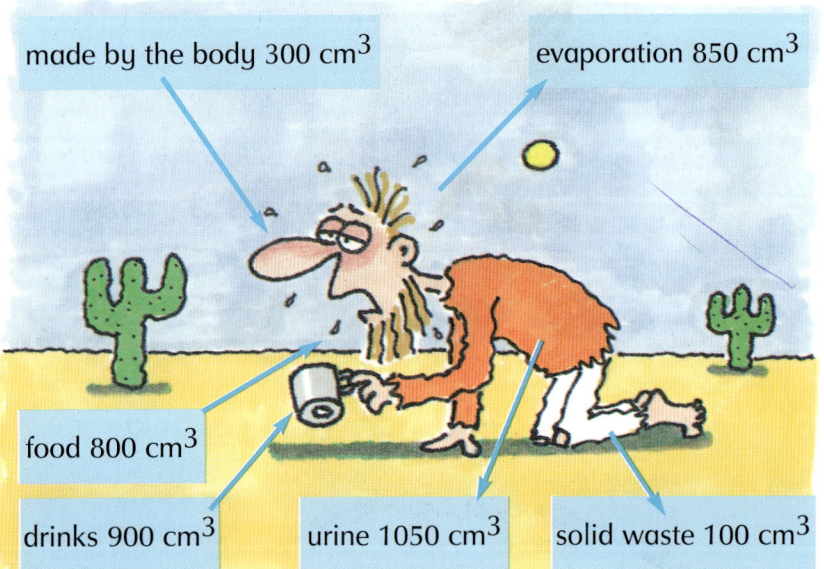

made by the body 300 cm^3

evaporation 850 cm^3

food 800 cm^3

drinks 900 cm^3

urine 1050 cm^3

solid waste 100 cm^3

We can only live for a few days without having a drink. 70% of our body is water. Water is needed to make blood and digestive juices. The **kidneys** control the amount of water in our body. The **bladder** stores urine until you go to the toilet.

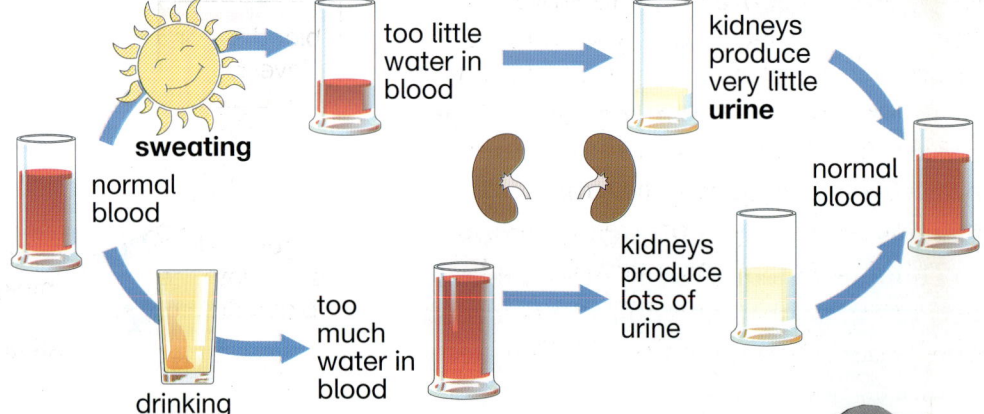

sweating

normal blood

too little water in blood

kidneys produce very little **urine**

normal blood

drinking

too much water in blood

kidneys produce lots of urine

1 Why is it important for people to drink water?
2 Which part of your body gets rid of extra water?
3 Why do you produce less urine on days when you do not have much to drink?
4 Work out how much water we take in each day and how much water we lose in a day.

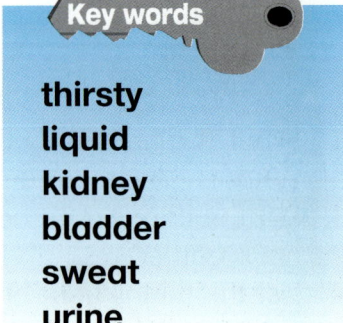

Key words

thirsty
liquid
kidney
bladder
sweat
urine

6.4 Controlling blood sugar

David was born with **diabetes**. He needs to take medicine every day to control his blood sugar levels. But it doesn't stop him climbing mountains on a school trip.

- *Who do you know who has diabetes?*
- *Is there anyone at school who has diabetes?*

A lot of the food we eat contains **sugar**. We use some sugar for energy straight away and store the rest for later. A chemical called **insulin** tells your body when there is spare sugar to store. People who do not make insulin have an illness called diabetes. They do not store extra sugar. Too much sugar stays in their blood. At other times they run out of sugar. They may feel faint, become unconscious or even go into a coma. Diabetic people need to **inject** themselves with insulin every day and be careful about the food they eat.

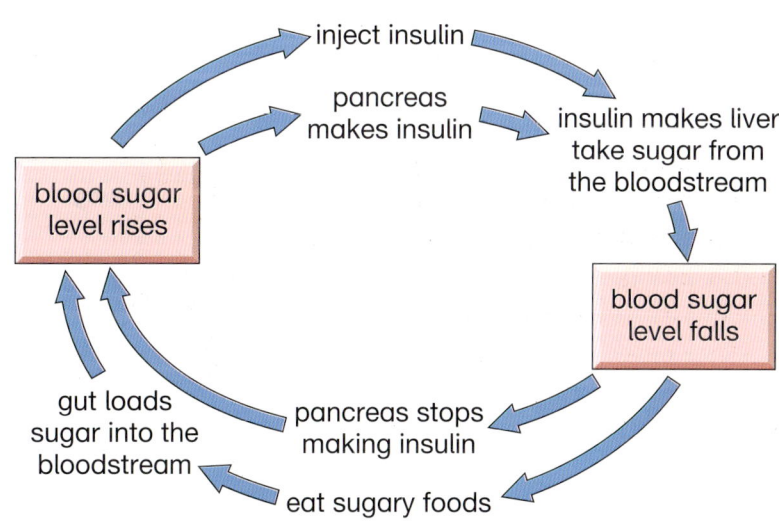

inject insulin

pancreas makes insulin

insulin makes liver take sugar from the bloodstream

blood sugar level rises

blood sugar level falls

gut loads sugar into the bloodstream

pancreas stops making insulin

eat sugary foods

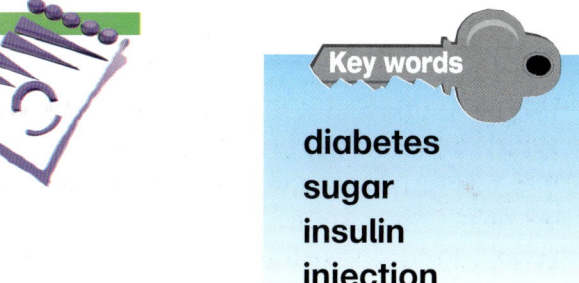

1 What do we use sugar for?
2 What happens to the sugar that we do not use straightaway?
3 What is diabetes?
4 How can diabetes be treated?
5 What could happen to a diabetic person who does not have his or her injections?

Key words

diabetes
sugar
insulin
injection

7 Acids and alkalis

7.1 Colouring clothes

The colour in these traditional clothes probably come from a plant **dye**. Different plants give different coloured dyes. Onion skins can dye white cotton cloth yellow. Your jeans were probably dyed with a dye called **indigo** which is also made from a plant.

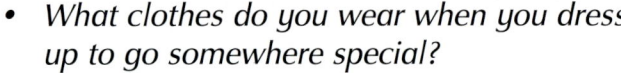

- *What clothes do you wear when you dress up to go somewhere special?*
- *What often happens when you wash a new pair of jeans?*
- *Have you ever bought pre-washed jeans? How do they look different?*

A dye is a substance that can change the colour of something else. As well as **natural** dyes such as onion skins, there are **artificial** dyes which tend to provide the bright colours in clothes.

Sometimes a substance is added to help the dye stick to the cloth. This is called a **mordant**. Here are some results obtained when cotton cloth was heated with water containing onion skins.

Temperature /°C	Onions skins and salt	Onions skins only
	Dye absorbed /%	Dye absorbed /%
60	15	8
80	20	10
100	25	12

1 What is a dye?
2 What do these results tell you about the best temperature for dyeing cloth?
3 What is a mordant?
4 Is salt a mordant for the dye in onion skins? Explain your answer.

7.2 Colouring food

Food is often coloured to make it look nicer. Food dyes can be natural dyes made from plant and animal materials or artificial dyes made from coal.

- *Why do you think people want food to look nicer?*
- *What taste would you expect green food to have? blue food? yellow food?*

Many Indian foods contain a natural dye called **turmeric**. Turmeric is made from the root of a plant. Turmeric changes colour in an **acid** or in an **alkali**. A dye which changes colour when it goes from an acid to an alkali is called an **indicator**.

In vinegar (an acid) turmeric is orangey-yellow. In washing powder (an alkali) turmeric turns pink. You can make indicator paper by dipping filter paper in turmeric and leaving it to dry. You can use this paper to test for acids and alkalis.

	Acid	Alkali
Taste	sour	
Litmus paper	pink	blue
Turmeric paper	orangey-yellow	pink
Universal indicator	orange	blue
Examples	vinegar lemon juice hydrochloric acid	washing soda sodium hydroxide many soaps

Litmus paper can also detect acids and alkalis. Litmus is a kind of moss which grows in cold parts of the world. The dye is extracted from the moss and soaked into filter paper.

1 What is an indicator?
2 What colour does litmus paper go in an acid?
3 What colour does litmus paper go in an alkali?
4 How could you use litmus paper to see which foods are acids?

7.3 Curing indigestion

Your stomach contains about one litre of hydrochloric acid. This breaks down your food. If there is too much acid in your stomach you will feel the pain called **indigestion**. Indigestion tablets contain a weak alkali called bicarbonate of soda. When acids and alkalis react together they form a **neutral** solution. The reaction is called **neutralisation**.

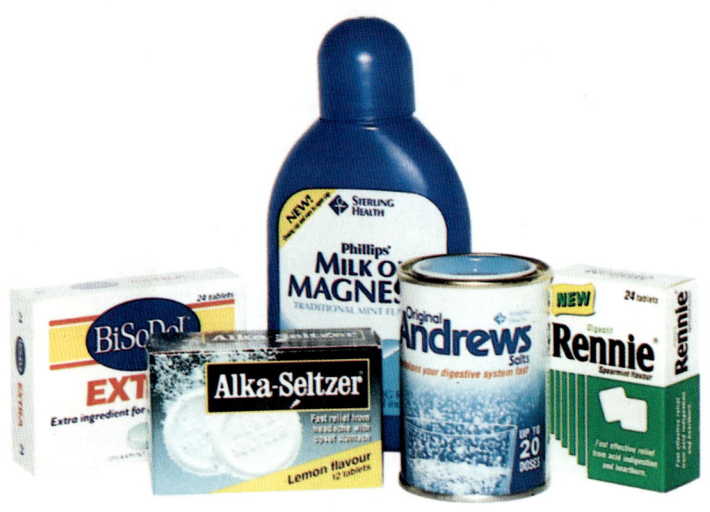

- *Do you get indigestion?*
- *What sorts of food give you indigestion?*
- *Do you take indigestion tablets or do you wait for the pain to go by itself?*

Neutralisation produces a solution which is not alkaline or acid. Pure water is neutral. It does not change the colour of litmus paper. So you could have blue and red litmus paper in the same solution at the same time.

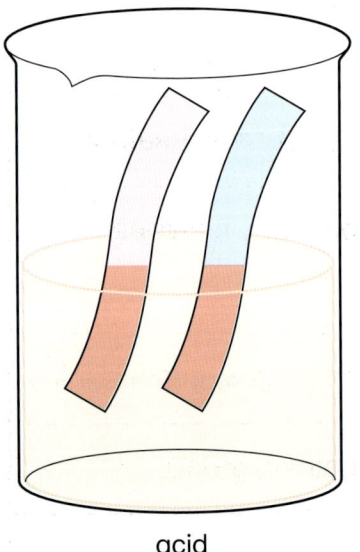

acid

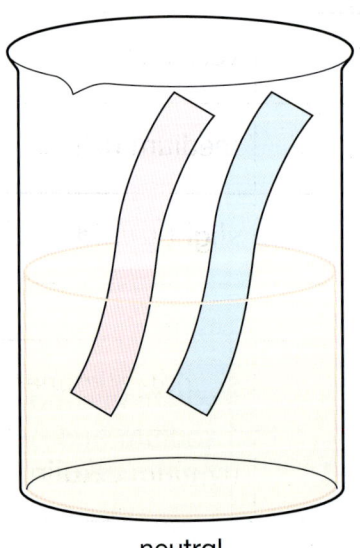

neutral

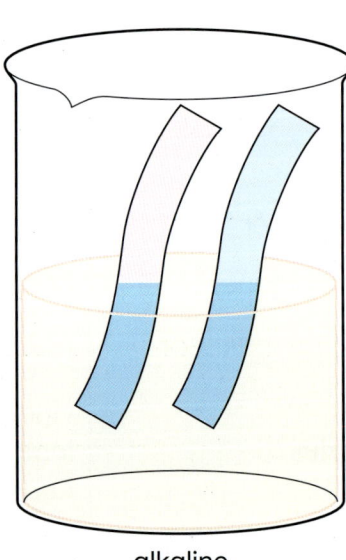

alkaline

1 What is a neutral solution? Give an example.
2 What colour does a piece of red litmus paper go in a neutral solution?
3 What happens when indigestion tablets are added to hydrochloric acid?
4 Why are indigestion tablets sometimes called antacids?
5 Plan an investigation to see which type of indigestion tablets is the best.

Key words

indigestion
neutral
neutralisation

7.4 Curing soil acidity

Gardeners have to make sure that the acidity of the soil is right. Most plants prefer the soil to be neutral or very slightly acidic. If the acidity of the soil is wrong the plant cannot grow.

- Have you tried to grow plants?
- What makes a plant grow well?
- How could you cure a soil that is too acid?

Some plants can grow in soil which would not suit other plants. The soil on moorlands is often very acidic. Heather is one of the few plants that grow well here.

Farmers need to know more than just whether a soil is acid. They need to know exactly how acid it is. Very acid soils are useless but slightly acid soils can grow some crops. They give the soil a score out of 13. This is called the **pH**. If the soil is too acid the farmer can add **lime**. But he must not add too much or he will make the soil too alkaline!

Universal indicator colour	pH	Acid or alkali	Plants that will grow
red	1 2	very acid	nothing grows
reddish orange	3 4	medium acid	blueberries, heather
yellowish orange	5 6	slightly acid	peanuts, potatoes
yellow	7	neutral	plums
green	8 9	slightly alkaline	cabbages, gooseberries
blue	10 11	medium alkaline	nothing grows
violet	12 13	very alkaline	nothing grows

1 Is a pH of 9 acidic, alkali or neutral?
2 What is the pH of neutral soil?
3 Some soil has a pH of 5. The farmer thinks the soil is too acid. What could he add to the soil to make it neutral?
4 What could the farmer grow in his soil if the pH was 6?

Key words

pH
lime

8 Cooking chemistry

8.1 Dissolving tastes

Some of the substances in this picture are solids
and some are liquids. Some of the solids will
dissolve in a liquid to make a **solution**. A
solution is made from two things: a
solute and a **solvent**. When you make
a cup of instant coffee you are making
a solution.

coffee powder + hot water → coffee
      solute   + solvent   → solution

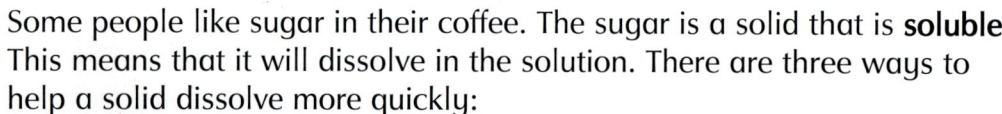

- *Think of the substances in your kitchen at home.
  Sort them into two groups: solids and liquids.*
- *Now sort the solids into soluble and insoluble ones.*

Some people like sugar in their coffee. The sugar is a solid that is **soluble**.
This means that it will dissolve in the solution. There are three ways to
help a solid dissolve more quickly:
1   stir the liquid
2   make the solid particles smaller
3   heat the liquid.
Some solutes do not dissolve in water. They are **insoluble** in water.
They may dissolve in another solvent. Chocolate is a good example:
it does not dissolve in water but does dissolve in cooking oil.

1   What is a solute? Name some.
2   What is a solvent? Name some.
3   Name the three ways that dissolving can be speeded up.
4   Plan an investigation to find out which of your answers to
    question 3 has the biggest effect.

Key words

dissolve
solution
solute
solvent
soluble
insoluble

8.2 Mixtures

This giant pizza is a complicated **mixture** of different substances – cheese, tomatoes, dough, bits of meat and so on. A mixture is something that contains two or more different **substances**.

- *How many ingredients are there in the pizza?*
- *How long do you think this pizza would take to cook?*
- *Are any of the **ingredients** solutions?*

Make a flat base with bread dough (flour, salt, yeast and water)

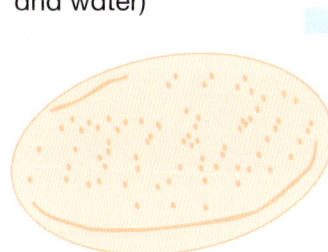

Add tomato paste

Add grated cheese

Add herbs and other flavours and cook

The salt is a pure substance: it contains only salt. You cannot **separate** it into two substances. All the other ingredients are mixtures.

Most of the things that we cook are made from mixtures. These mixtures often need to be mixed together to make a mixture of mixtures. The chemistry of cooking is very complicated!

1 What is the difference between a pure substance and a mixture?
2 List ten mixtures from your kitchen at home. Now try to list five pure substances.
3 Which are more common: mixtures or pure substances?
4 Is bread dough a mixture or a pure substance? Why?

Key words

mixture
substance
ingredients
separate

33

8.3 Compounds and elements

Everything is made of **atoms**. There are about one hundred different types of atom. In different combinations they make up every single thing in the universe. The air we breathe contains oxygen. Oxygen is made only of oxygen atoms. Since oxygen contains only one type of atom we say oxygen is an **element**.

- *How many atoms do you think would fit on the head of a pin?*
- *If an atom was the size of a tennis ball how tall do you think you would be on the same scale?*

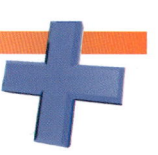

The air also contains carbon dioxide. Carbon dioxide contains oxygen and carbon atoms. The oxygen and carbon atoms have joined together to make a **compound**. Pure carbon dioxide contains only carbon dioxide. But we know that carbon dioxide contains two sorts of atoms: carbon and oxygen.

The purple gas in this tube is iodine. It is another element. It only contains iodine atoms.

Coffee is a mixture of compounds: it contains lots of different substances. Each of these substances has more than one type of atom.

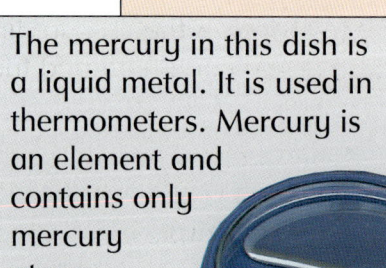

Sugar is a pure compound. Sugar contains atoms of hydrogen, oxygen and carbon.

The mercury in this dish is a liquid metal. It is used in thermometers. Mercury is an element and contains only mercury atoms.

1 What is an atom?
2 About how many different types of atom are there?
3 What is an element?
4 List three compounds.
5 What is the difference between a compound and a mixture?

Key words

atom
element
compound

8.4 Chemical reactions

Raw cake mixture is not the same as a cooked cake. When we cook the mixture it forms a new substance. This change is called a **chemical reaction**. A chemical reaction **changes** the starting mixture into something completely new. You cannot break up the cake into crumbs to get the original ingredients back.

- *Do you still have a birthday cake every year?*
- *Think of five permanent changes from your kitchen.*
- *How is cooking a cake different from freezing ice cream?*

In every chemical reaction two things happen:

1 a new substance forms
2 the change is difficult to undo.

Both of these changes happen because the atoms in the mixture link up in different ways. The atoms are the same at the beginning and the end – but they are arranged in a different way.

Self-raising flour is used in many cake recipes. Self-raising flour contains a chemical which gives out carbon dioxide when it gets hot. The carbon dioxide makes the cake rise. A good sponge is full of carbon dioxide **gas**! We can show the reaction as a word equation:

baking powder → carbon dioxide + sodium carbonate

1 What two things happen when chemicals react together?
2 What happens to the atoms in a chemical reaction?
3 List examples of chemical reactions that occur in your kitchen.
4 Plan an investigation to find out which brand of baking powder is best.

Key words

chemical reaction
change
gas

9 Cleaning up

9.1 Graffiti

The people who painted this wall probably wanted their message to last forever. They will be disappointed. A solvent could wash the paint from the wall. A solvent is a liquid that dissolves something. The solvent dissolves the paint so that it can be wiped away from the wall.

- *Many people think graffiti is always bad and should be cleaned away. Do you agree?*

The thing the solvent dissolves is called the solute. When the solute dissolves in the solvent the liquid is called a solution. When the solvent dries out the solute is left behind.

The person who has to clean the wall will need to choose the right solvent for the job. One solvent cannot dissolve everything. Many solvents are powerful chemicals that can be dangerous.

Stain	Solvent	Dangers
gloss paints	**white spirit** **turps**	burns very easily
nail varnish	nail varnish remover	
waxes and grease	**paraffin** **meths**	**poisonous** fumes

1 What is a solvent?
2 Which solvent could I use to clean gloss paint from paintbrushes?
3 People warm water to help it to dissolve salt more easily. Why is this a very dangerous thing to do with meths?
4 Plan an investigation to find out which solvent will clean ink from white cloth.

Key words

white spirit
turps
paraffin
meths
poisonous

9.2 Soaps and detergents

Imagine the team cleaning their kit using only water. If they tried to dissolve the mud and wash it away it would take a very long time! They will have to use **soap** or a **detergent**.

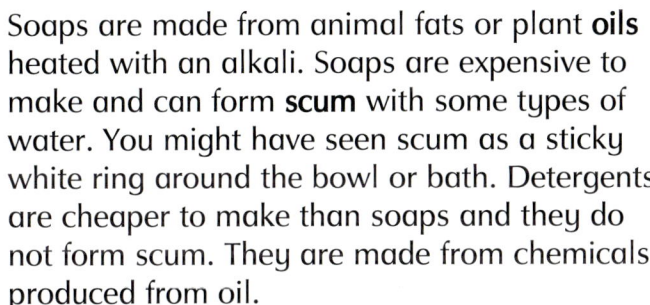

- *Have you ever been as dirty as this?*
- *Did you have to wash the kit yourself?*
- *What can go wrong if you wash something with the wrong soap or detergent?*

Soaps are made from animal fats or plant **oils** heated with an alkali. Soaps are expensive to make and can form **scum** with some types of water. You might have seen scum as a sticky white ring around the bowl or bath. Detergents are cheaper to make than soaps and they do not form scum. They are made from chemicals produced from oil.

The table shows the time needed to clean a blackcurrant stain using Scotto soap powder at different temperatures.

Temperature / °C	Time needed / mins
40	30
50	20
60	10

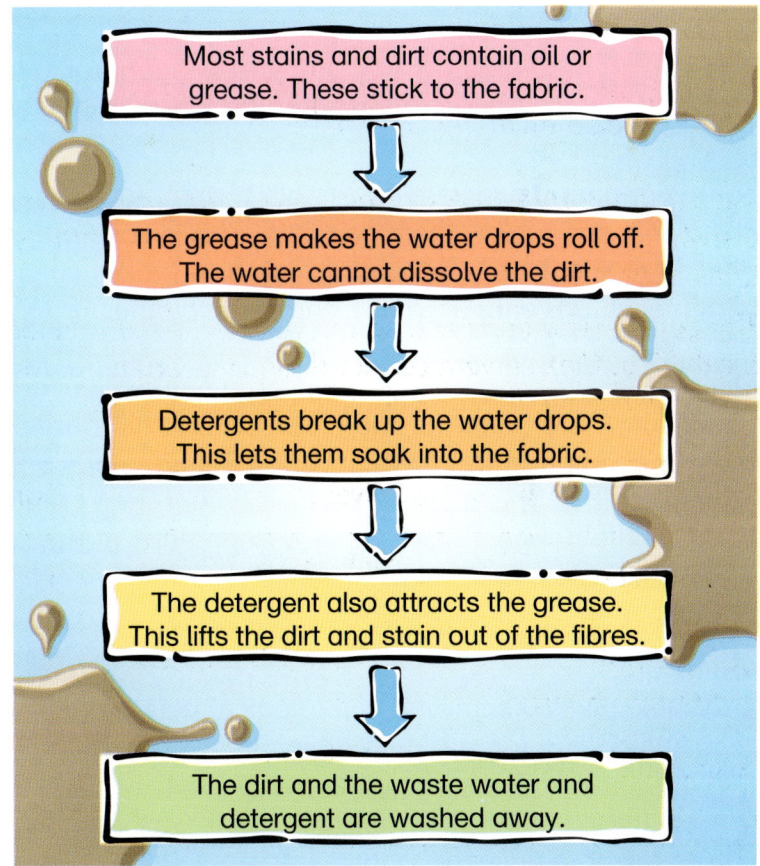

Most stains and dirt contain oil or grease. These stick to the fabric.

The grease makes the water drops roll off. The water cannot dissolve the dirt.

Detergents break up the water drops. This lets them soak into the fabric.

The detergent also attracts the grease. This lifts the dirt and stain out of the fibres.

The dirt and the waste water and detergent are washed away.

1 Give two differences between soap and detergents.
2 Explain how detergents work.
3 Draw a bar chart to show the time needed to clean blackcurrant juice using Scotto at different temperatures.
4 Why don't we always boil clothes when we wash them to make sure they are clean?

Key words

soap
detergent
oil
scum

9.3 Those difficult stains

Babies are always messy!
Some of the foods will **stain** his
clothes. You could clean the stains
with very hot water but this would spoil
the clothes. Some washing powders
contain chemicals called **enzymes**.
Enzymes can digest food stains in the
same way that they digest the foods
that you eat. Washing powders with
enzymes are often called
biological powders.

- *Do you use biological washing powder at home?*
- *Have you ever had clothes spoilt by washing powder?*

Some washing powders also contain **bleach**. Bleach reacts with coloured
chemicals and makes them go white. This means any stains are more
difficult to see. Unfortunately, bleach also makes colours fade. Powders
with bleach are only used to clean white clothes or fabrics.

Delicate fabrics like wool or silk need a gentle cleaner. Soap flakes or
liquid soaps are better for these fabrics. Not all fabrics can be cleaned at
the same temperature. A white cotton T-shirt needs a hot wash. Jeans are
not **colour fast**. This means that their colour may come out in the wash.
What would happen if you washed your jeans at the same time as a
white T-shirt?

1 What special chemicals do biological detergents
   contain?
2 Design the outside of a detergent packet.
3 Sort these clothes into four groups for washing: silk shirt,
   tea towel, blue jeans, baby's bib, pure wool sweater,
   cotton shirt with oil on, patterned sheet, dish cloth,
   tablecloth, white towel, baby's vest, football kit.

Gentle soap flakes	Normal detergent	Biological detergent	Powder with bleach

9.4 Cleaning the bath

We use soaps in the bath, not detergents. This means we often get a tidemark around the bath. This mark is made of scum. It forms when soaps react with chemicals dissolved in the tap water. **Hard water** contains lots of dissolved substances. It produces a lot of scum and it is difficult to make it produce bubbles. **Soft water** contains less dissolved substances. It produces very little scum and needs very little soap.

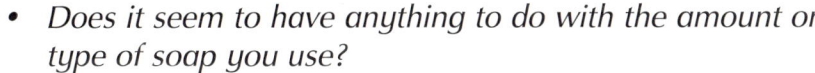

- *Do you notice a lot of scum after you have had a bath?*
- *Does it seem to have anything to do with the amount or type of soap you use?*
- *Bath oils often make scum worse. Suggest a reason why.*
- *Do you live in a hard or soft water area?*

Abrasive cleaners
These are usually powders or creams that contain small bits of grit which scratch the scum off the surface of the bath. Abrasive cleaners are suitable for china and porcelain.

Non-abrasive cleaners
These contain solvents that dissolve the grease in the scum. They are usually sold as creams or liquids. They are more expensive than the abrasive cleaners but they do not scratch soft surfaces. They are used to clean plastic and glass.

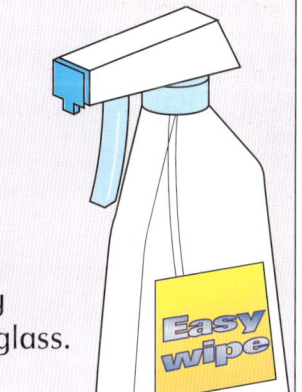

In hard water areas **limescale** forms on taps and baths. Limescale is made of crystals left behind when water evaporates. You can use non-abrasive cleaners to get rid of limescale. Acids in the cleaners dissolve the crystals.

1 Explain the difference between hard and soft water.
2 What does abrasive mean?
3 How does a non-abrasive cream clean scum from a bath?
4 Why is it a bad idea to use an abrasive cleaner on a plastic bath?
5 What causes limescale?

Key words

hard water
soft water
abrasive
non-abrasive
limescale

10 Precious metals

10.1 King Tut's curse

The Egyptian King Tutankhamen died over three thousand years ago. His solid gold death mask contained 200 kilograms of gold and precious stones.

When Howard Carter found the tomb in 1922, the death mask was still as good as the day it was made. The tomb contained a curse to protect the dead king's treasures from thieves. Some of the people who discovered the mask died soon afterwards. Some people still believe that anybody who owns the mask is cursed.

- *Do you think the mask is cursed?*
- *Who do you think should own the mask – the person who discovered it or the country where it was found? Why?*

Gold does not react easily with other chemicals. This is why King Tut's mask lasted over 3000 years. When a metal **reacts** with the air it changes into a new chemical. Scientists say it has **corroded**. An **iron** gate rusts badly in a few years, unless it is painted so air cannot reach the iron surface.

In 1996, one grammè of gold cost roughly £12. That's a speck of gold that is roughly the same size as a grain of sand! At the same time, all the iron in a car engine cost less than £20.

Gold was used in the earliest times, long before bronze or iron. You can find tiny specks of pure gold in the rivers and streams of North Wales.

Gold is just like any other metal. It is shiny and can be beaten into flat sheets. Heat and electricity pass easily through gold and iron.

1 List four things that are true about gold and iron.
2 Give two differences between gold and iron.
3 Why do you think that metal workers used gold long before bronze or iron?
4 How would you recognise tiny fragments of gold in river gravel?
5 How much is the gold in Tutankhamen's mask worth?

Key words

react
corrode
iron

10.2 Gold jewellery

Gold is still used for jewellery. Pure gold jewellery is soft and very expensive. Jewellers often mix gold with other metals like **nickel** or **silver** to make a harder, cheaper **alloy**. An alloy is a mixture of two pure metals.

- *Why is pure gold sometimes less useful than an alloy of gold and silver?*
- *Pure gold jewellery costs more than alloy jewellery. Do you think it is worth the extra money? Why?*

		Hallmark	
22 carat		👑	916
18 carat		👑	750
14 carat		👑	585
9 carat		👑	375

Amount of gold in 1000 g of the alloy

People would try to bite a gold coin – if it was soft enough to dent, it was pure gold.

Gold is easy to hammer into flat sheets called **gold leaf**. Other metals are more difficult to flatten in this way.

Gold is a very heavy metal.

Acid reacts with many metals – but not gold.

1 What is an alloy?
2 Why is an alloy sometimes better for jewellery than pure gold?
3 Why is an 18-carat gold ring cheaper than a 22-carat ring?
4 Plan a test to discover if a gold-coloured substance brought into the laboratory is pure gold or not.

Key words

nickel
silver
alloy
gold leaf

41

10.3 Silver plate

Manufacturers try to make cheap jewellery look **expensive**. They coat the jewellery with metals like silver, gold or platinum. There are two ways to do this.

One is to dip the piece in molten gold, silver or platinum, pull it out and let a layer of precious metal solidify onto the cheap metal jewellery. This gives a thick and uneven **coating** of precious metal. A better method is **electroplating**. Electroplating produces a thin, even coating of the precious metal.

- *Have you got anything made of silver?*
- *How can you tell if it is pure silver or silver plated?*

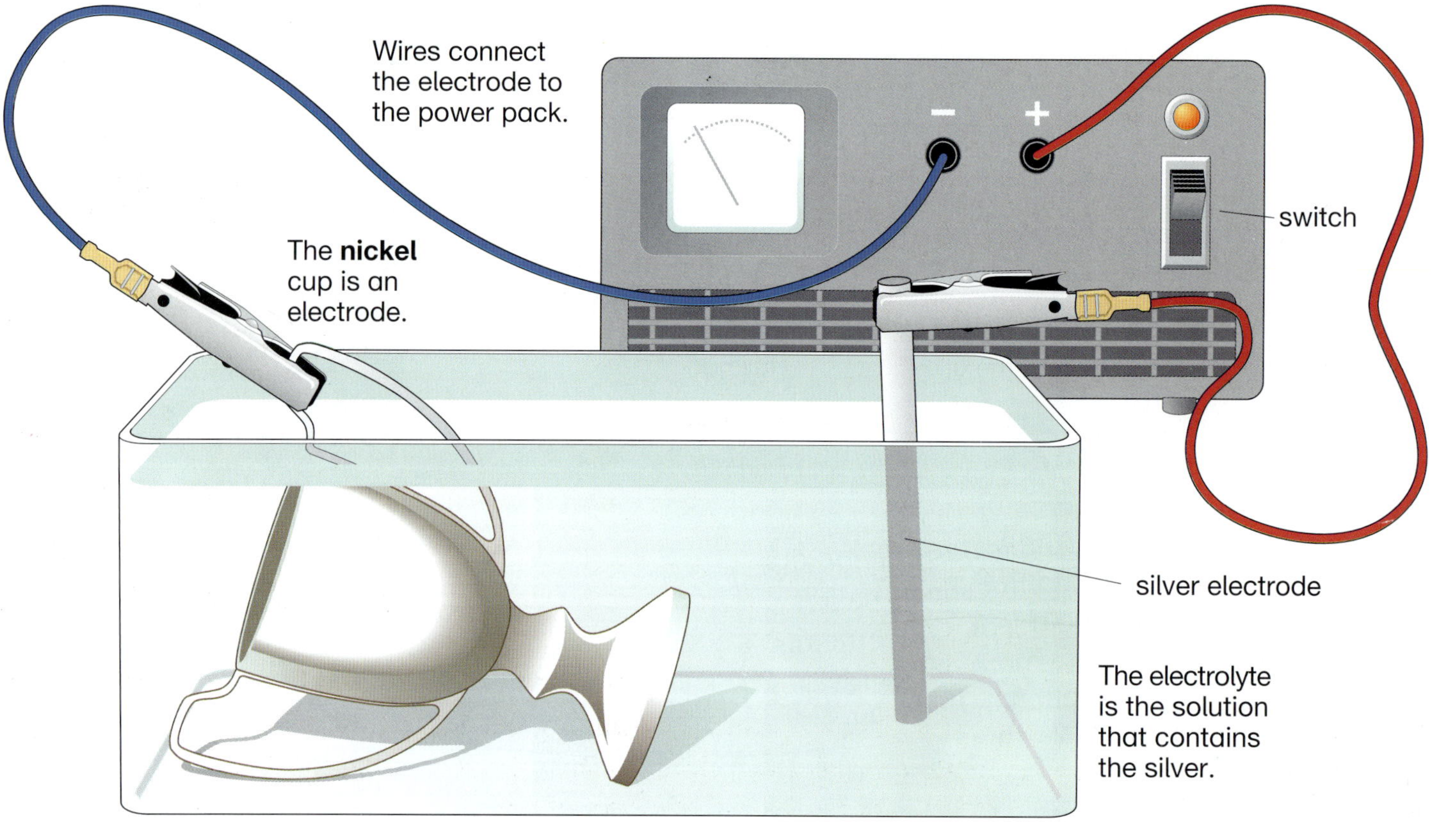

Wires connect the electrode to the power pack.

The **nickel** cup is an electrode.

switch

silver electrode

The electrolyte is the solution that contains the silver.

1 Give two ways to coat a metal ring with gold.
2 What is an electrode?
3 Is the nickel cup the positive or negative electrode?
4 What is the electrolyte in this experiment?
5 What might affect the thickness of silver on the ring?
6 Why is a thin, even coating of precious metal better than a thick, uneven coating?

Key words

expensive
coating
electroplating
nickel

10.4 Recycling silver

Chemicals containing silver react to light. This makes silver compounds very useful in photographs. A black and white photograph is a pattern of silver grains spread across a white paper background. Laboratories which develop films produce large amounts of waste silver. This silver stays in the liquid left over after developing the film. Silver is expensive so laboratories try to get the silver back from these waste solutions.

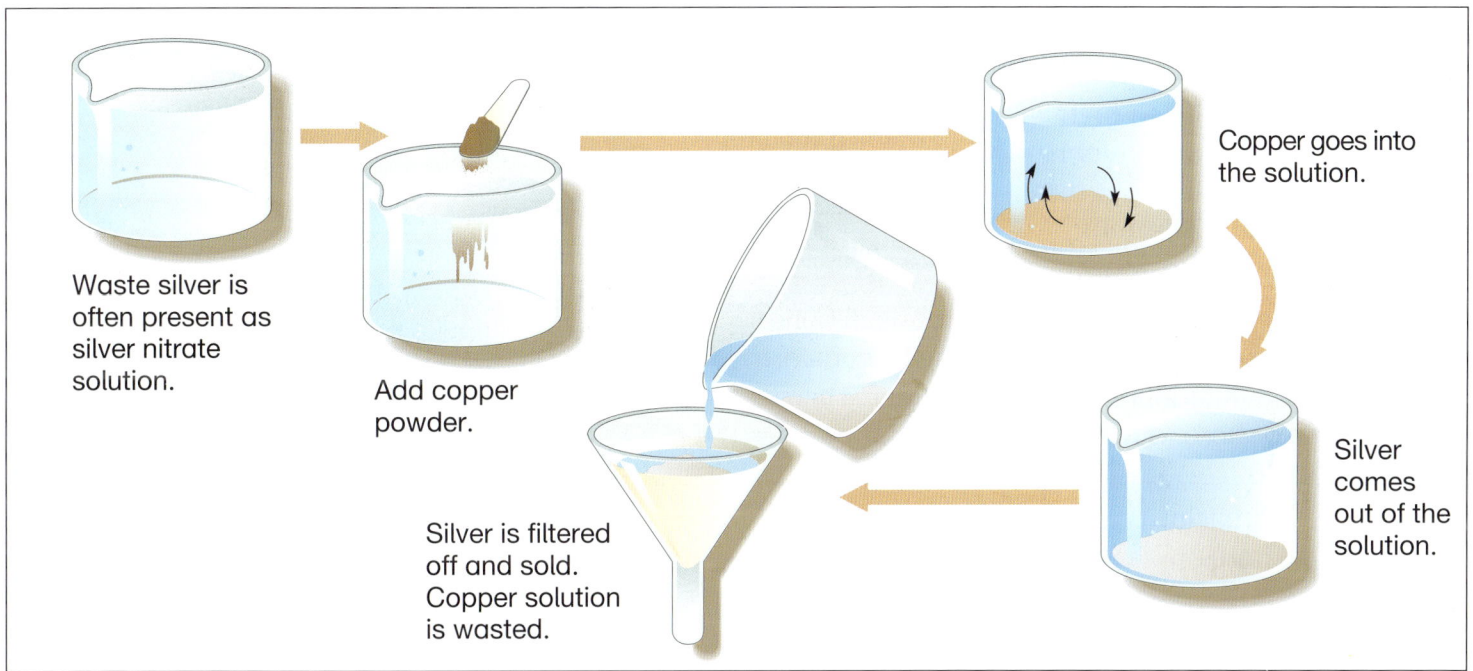

Waste silver is often present as silver nitrate solution.

Add copper powder.

Copper goes into the solution.

Silver comes out of the solution.

Silver is filtered off and sold. Copper solution is wasted.

This is an example of a **displacement reaction**. The silver displaces the **copper** from the solution. You can tell if a metal will displace silver from the solution by looking at its **reactivity**. Reactivity is a measure of how easily a metal reacts with other chemicals. A metal that is more reactive than silver will displace it from the solution.

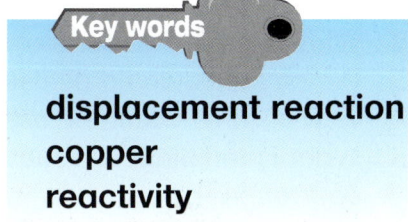

1 Why is silver used in photographic film?
2 What does reactive mean?
3 What is a displacement reaction?
4 Why don't the laboratories worry about wasting copper?
5 Plan an investigation to find out which metals can displace copper from a solution.

Key words

displacement reaction
copper
reactivity

11 Clothes

11.1 Fibres and fabrics

- *Have you ever been camping?*
- *Did you buy or borrow special clothes for the trip?*
- *What were these clothes made from?*
- *Did the clothes do what they were supposed to do? Were they waterproof and warm?*

Clothes are made from different **fabrics**. Some of the fabrics are made from **threads** woven together. Others seem to be a sort of plastic. The woven fabrics are thicker and warmer but the plastic ones look waterproof. Which ones are the best?

> The mist came down very quickly. In minutes I was cold and wet. If I hadn't been wearing the proper clothes I could have been in real danger.

In woven fabrics the properties of the fabric depend on the thread used to make them. These threads are made of even thinner **fibres** twisted together. There are two sorts of fibre: **natural** and **artificial**.

Natural fibres	Artificial fibres
Silk, wool, mohair, cotton, linen	Acrylic, lycra, polyester

Fibre	Properties
Acrylic	Strong, soft, warm
Lycra	Very stretchy, springs back into shape
Nylon	Stretchy, strong, tough
Polyester	Quick drying, tough, blends well with other fibres
Wool	Stretchy, soft, warm
Cotton	Stretchy, can be cool or warm, absorbs water

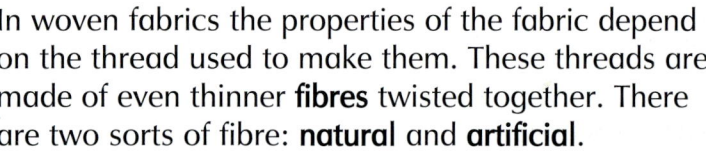

1. What is a fabric?
2. What are the two sorts of fibre?
3. Which fibres would be useful for a T-shirt or vest? Why?
4. Which fibres would be useful for waterproof leggings? Why?
5. What is the biggest disadvantage of wool? Explain your answer.
6. Plan an investigation to find the strength of a thread.

Key words

fabric
thread
fibre
natural
artificial

11.2 Different fabrics

Some jackets are made from a special 'breathing' fabric. This fabric keeps you warm and keeps water out but lets sweat pass from your body to the outside. This one-way fabric is very expensive. The jacket costs almost twice as much as ordinary ones.

- *What is the most expensive set of clothes you have ever bought?*
- *Why were they so expensive?*
- *Were they worth the money? Explain your answer.*
- *When might you want a jacket made from fabric which 'breathes'?*

Sometimes the way the cloth is made makes it very expensive. The **weave** affects the way the fabric behaves when it is is used. Tight weaves tend to be **waterproof** and **windproof**. Open weaves stretch easily and can trap air layers which help to keep people warm. Different **yarns** cost different amounts of money. Sometimes people are willing to pay more for a designer label or a different colour.

| A student weighed a beaker with fabric fixed on the top. | She poured 10 ml of water slowly over the fabric. Some of the water went through into the beaker. | She weighed the beaker and the wet fabric. Then she tried again with four different fabrics. |

Fabric	Dry fabric + beaker /g	Wet fabric + beaker /g	Weight water in the beaker /g
Cotton	95	102	7
Waxed cotton	93	95	2
Nylon	90	94	4
Silicone nylon	98	99	1
Rubber	94	94	0

1 Draw a bar chart of the the weight of water against the fabric.
2 Why use 10 ml of water for each fabric?
3 Which fabric is completely waterproof?
4 Which fabrics are only **water-resistant**?

Key words

weave
waterproof
windproof
yarn
water-resistant

11.3 Special fabrics

Firefighters depend on their clothes to save their lives. The fabrics of their clothes reflect heat. They are treated with special chemicals so that they do not burn easily. Even so, fighting fires is a dangerous and difficult job.

- *List the ways the clothes protect the firefighter in the photograph.*
- *Why do you think the clothes are silver-coloured?*
- *Would you feel safe in clothes like these?*
- *Would you work as a firefighter?*

All fabrics burn if they become hot enough. However, we can make it more difficult for them to catch fire. Some chemicals seem to reduce the chance of a fabric catching fire. We call fabrics that have been treated with these chemicals **flameproof**.

A student tested some fabrics to see how they burned before and after flameproofing. He used three equal pieces from each fabric and treated them with flameproofing chemicals.

untreated

treated with **alum**

treated with **borax**

He heated the fabrics with a bunsen burner. He timed how long before each fabric caught fire.

Fabric	Untreated /sec.	Alum /sec.	Borax /sec.
Cotton	30	35	39
Nylon	23	30	36
Wool	25	32	37

1 What does flameproof mean?
2 Which fabric would be most suitable for making a flameproof coat? Why?
3 Which treatment would you use? Why?
4 Firefighters' uniforms need to be washed regularly. How might this affect their flameproof coating?

Key words

flameproof
alum
borax

11.4 Footwear

Hill walkers need **footwear** for their trip. Some people want walking boots to protect their feet from sharp stones. They must be strong, comfortable and have soles with a good **grip**. Some want comfortable, lightweight trainers with **non-slip** soles.

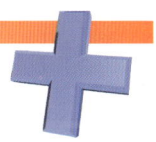

- *How many sorts of shoes do you have?*
- *Which are your most comfortable pair?*
- *What makes them the most comfortable?*
- *What is most important to you when you choose a new pair of trainers?*

Walking boot

tongue – is fixed to the outside of the boot on each side so that no water can get through to the person's feet.

laces – usually very long so that the boot can be tied on tightly around the foot and ankle.

Trainer

eyeholes – must not tear when the laces are pulled tight. They can be strengthened with metal rings or extra layers of fabric or leather.

waterproof uppers – often made from fabric or leather.

uppers – must be waterproof. Sometimes the uppers have special one-way material patches which let sweat out from the shoe but keep water outside.

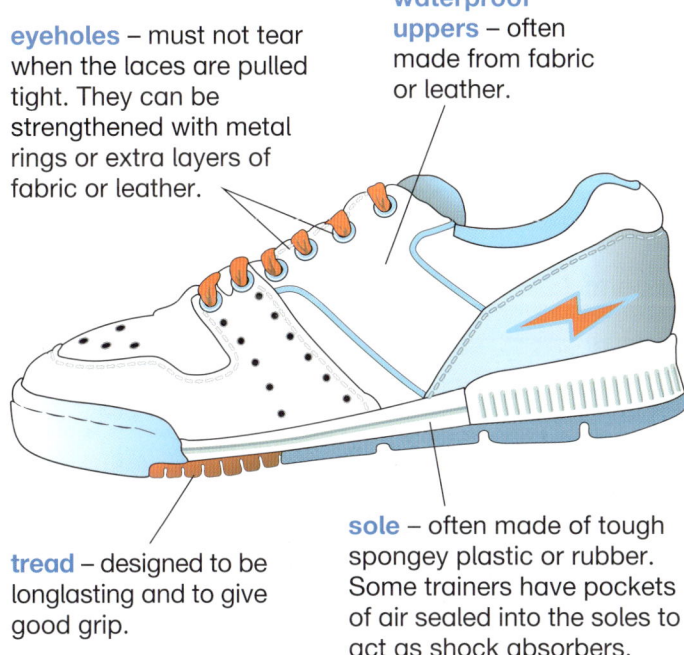

tread – designed to be longlasting and to give good grip.

sole – often made of tough spongey plastic or rubber. Some trainers have pockets of air sealed into the soles to act as shock absorbers.

heel – often built up to support the ankle.

sole – tough, and also soft and spongey to absorb shocks.

tread – usually a deep tread to give a grip on slippery surfaces.

1. What are the soles of the trainers and walking boots made from? Why?
2. List the materials used to make the uppers of the trainers.
3. Many shoes have rubber or plastic soles. Explain why.
4. List the parts of the walking boots. Suggest a material you could use to make each part.

Key words

footwear
grip
non-slip

12 Freezing and boiling

12.1 Melt down

Look at all this solid water! It is exactly the same chemical as the water in your tap. So why does it look so different? Ice is a different physical state to **liquid** water. Clouds are another physical state of water. Some scientists think that the Earth's temperature is rising slightly. They call this **global warming**. This may **melt** some of this ice.

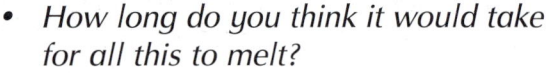

- *How long do you think it would take for all this to melt?*
- *Where would the water go when it melts?*
- *What effects might this have?*

Water is difficult to move around. Some parts of the world need a supply of fresh water for crops – and for people. Maybe we could tow an iceberg to these places and let the ice melt to make water?

The temperature at which a **solid** melts is called its **melting point**.

Substance	Melting point / °C
tungsten (used in light bulb filaments)	3400
table salt (sodium chloride)	801
lead	327
ice	0
mercury (used in some thermometers)	-39
oxygen (gas at room temperature)	-218

melts

ice → add heat → liquid water

freezes

1 What does the word melt mean?
2 How can we make something melt?
3 Which substance in the table has the highest melting point?
4 Which substance in the table melts at the lowest temperature?
5 Plan an investigation to find out how quickly an ice cube melts in cold and warm water.

Key words

liquid
global warming
melt
solid
melting point

12.2 Freeze frame

There's no problem with water freezing in pipes. But look what happens when the ice thaws! When water **freezes** the ice **expands** – it takes up more room than the water did. The pipes split. When the ice thaws again the water leaks out of the broken pipes.

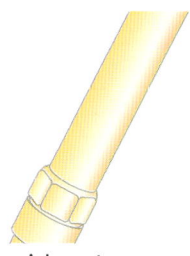
Liquid water runs through the pipe.

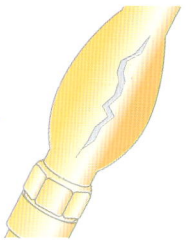
The water freezes and expands, causing the pipe to split.

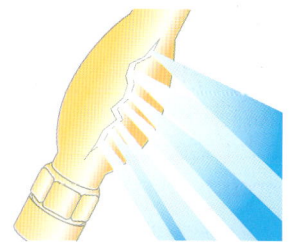
The ice melts and the water leaks through the split in the pipe.

- *Have you ever seen any burst pipes?*
- *What damage did the water do?*
- *Where are the mains water pipes at your home?*
- *Are they likely to freeze in very cold weather?*
- *How can you stop this happening?*

Antifreeze contains a chemical which does not freeze until the temperature drops to -16°C. The temperature at which something freezes is called its **freezing point**.

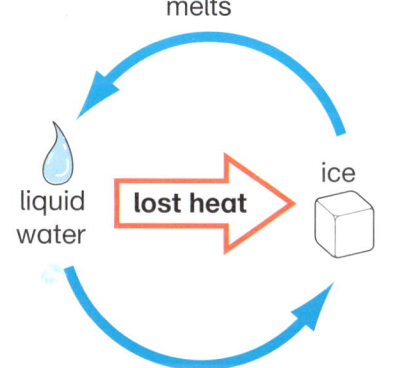

melts

liquid water **lost heat** ice

freezes

I develop new ice creams and lollies for a food company. I want to increase the amount of fruit juice in our lollies to make them taste better. I have tried some investigations with water and orange juice.

Mixture	Freezing point /°C
all water	0
50% juice	-2
75% juice	-4
all juice	?

1 What do we mean when we say something freezes?
2 What temperature does water freeze at?
3 What pattern can you see in the above results?
4 What temperature do you think pure orange juice might freeze at?

12.3 Runny or not?

Some things are not quite solid, but they are not quite liquid either. Is soft scoop ice cream a solid or a thick liquid? We call the runniness of liquids their **viscosity**. Thick liquids like honey have a high viscosity. Runny liquids like water have a low viscosity.

- *List as many things as possible that are not quite liquids or solids.*
- *How can you make a thick liquid into a solid?*
- *How can you make a thick liquid more runny?*

Oil in car engines helps the moving parts to slide over each other more easily. This reduces the **friction** and so keeps them from wearing out. The oil **lubricates** the engine. Cold oil is quite thick and takes a long time to coat all the moving parts. Hot oil can be so thin it runs off the parts too quickly. This means the parts are not protected. Oil companies test their oils so that they have the best viscosity – not too thick when cold and not too runny when hot.

1 What does viscosity mean?
2 Why is oil used in car engines?
3 What is the problem with thick cold oil?
4 How could you measure the viscosity of chocolate spread?
5 How could you find out how **temperature** changes the vicosity of chocolate spread?

Key words

viscosity
friction
lubricate
temperature

12.4 Where's it gone?

There is a village in Chile between the sea and the hills. It does not get a lot of rain. They do get a lot of fog. The villagers collect the fog and change it into liquid water. They use a large sheet to cool and collect the fog. The man in the photo is mending the sheet.

There is water in the form of a gas in the air. We call it **water vapour**. We can't see it. It is only when we cool the air down that we can see droplets of liquid water.

- *How do you think they can collect the fog?*
- *How can we cool the air down?*
- *There is water vapour in our breath. How can we show this?*

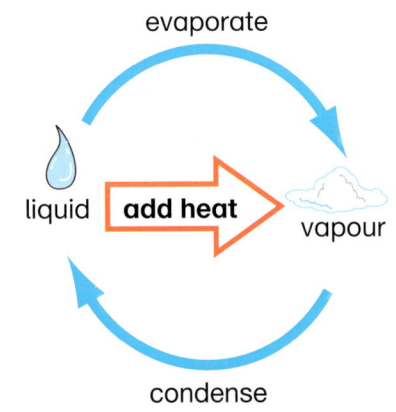

When gases get very cold they **condense** to make a liquid. Nitrogen is a gas at room temperature. If it is made very cold it can be kept as a liquid. Liquids **evaporate** when they change to a gas or vapour. So a puddle in the road dries out. We can heat the liquid to speed up evaporation. So the Sun shining on the road dries the puddle out more quickly. When a liquid changes state and becomes a gas or vapour we say it is boiling. The temperature at which a liquid boils is called its **boiling point**.

Substance	Boiling point /°C
pure water	100
methylated spirit	80
gold	3080

evaporate

liquid → add heat → vapour

condense

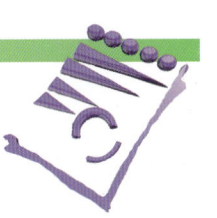

1 What happens when we cool a gas or vapour?
2 What do we mean by evaporation?
3 How can you speed up the evaporation of liquids?
4 What is the boiling point of pure water?

Key words

water vapour
condense
evaporate
boiling point

13 Lighting up

13.1 Switching off

During a power workers' strike all the lights in New York City failed. Over the next 12 hours murders doubled and thefts from shops and homes increased by ten times. Nine months later the birth rate trebled.

- List the things you use electricity for at home.
- What would happen if the power failed?
- Have you ever been in a power cut?
- What might you have used for light in the days before there was electricity?

Electricity needs a power supply and a complete **circuit** to flow. If the power fails or if there are any gaps in the circuit, the flow stops. We can tell when electricity flows through something because something happens. For example, a **bulb** lights up, a motor turns or an electric bell rings. A **switch** is a way to control electricity flow by opening and closing gaps in the circuit.

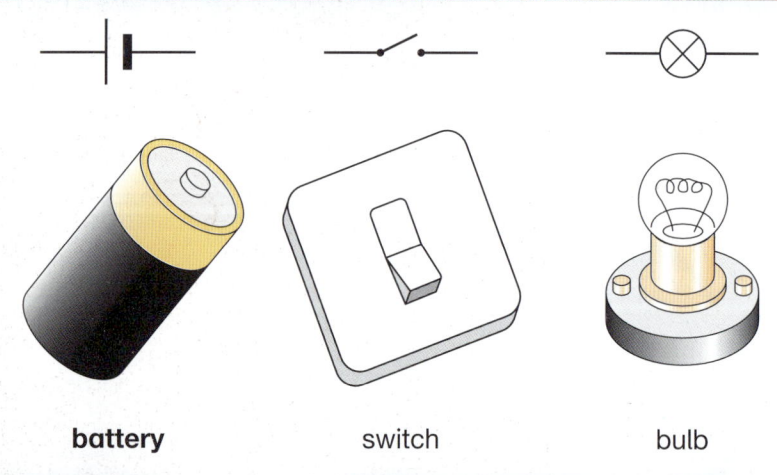

| battery | switch | bulb |

1. Make a list of six things that use electricity.
2. For each one suggest what you would have used before the days of electricity.
3. Draw a picture to show how you would use a battery and wires to light a bulb.
4. Put in a switch so you can turn the bulb on and off.
5. Draw a circuit diagram of your circuit using the correct **symbols**.

Key words

circuit
bulb
switch
battery
symbol

13.2 Conductors and insulators

- *How could you find out whether a material is a conductor or an insulator?*
- *Why is the wire in the drawing above covered in plastic?*
- *What other materials could you use for the electricity to go through?*

Electricity is useful – but dangerous. We need to be able to control where it goes. **Insulators** are things that stop electricity flowing. **Conductors** let electricity through.

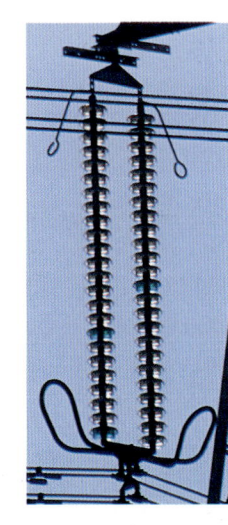

The wires on a **pylon** do not have insulation. Air is a very poor conductor so elecricity cannot pass to the ground. Metal is a very good conductor. Special glass or pottery disks keep the wire away from the metal pylon. These disks are called insulators.

1 What is a conductor?
2 List three materials that conduct electricity.
3 What is an insulator?
4 List three electrical insulators.

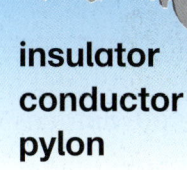

Key words

insulator
conductor
pylon

53

13.3 Bright lights

We use electricity to give us light. Sometimes it does not matter if a light goes out because the circuit breaks. Sometimes it could be very dangerous. Some circuits make failures less likely and so may be safer.

- *Where would a failure of the light be very dangerous? Why?*
- *Which is the brightest bulb in your home?*
- *Which room in your home has the best light? Why?*

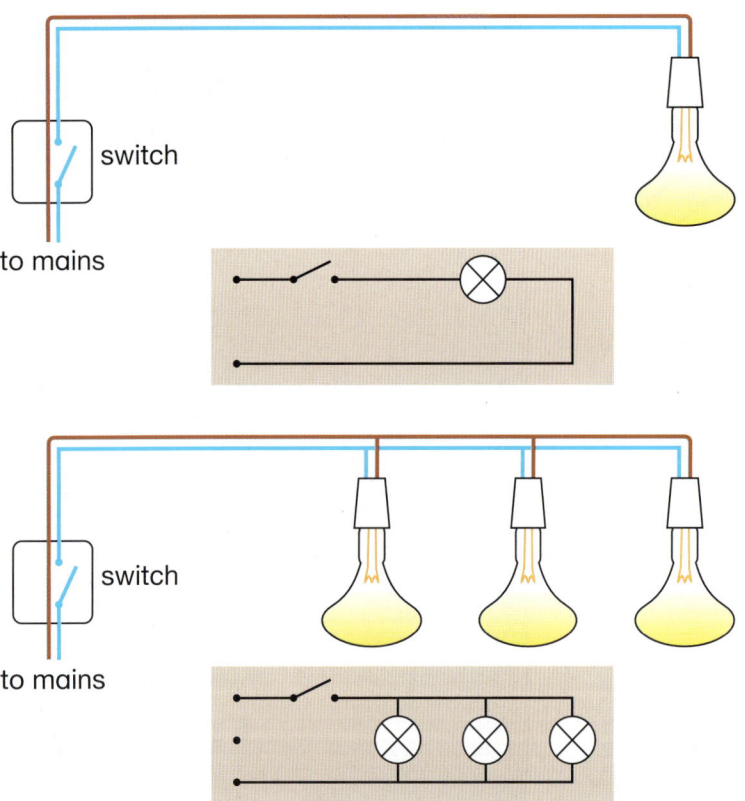

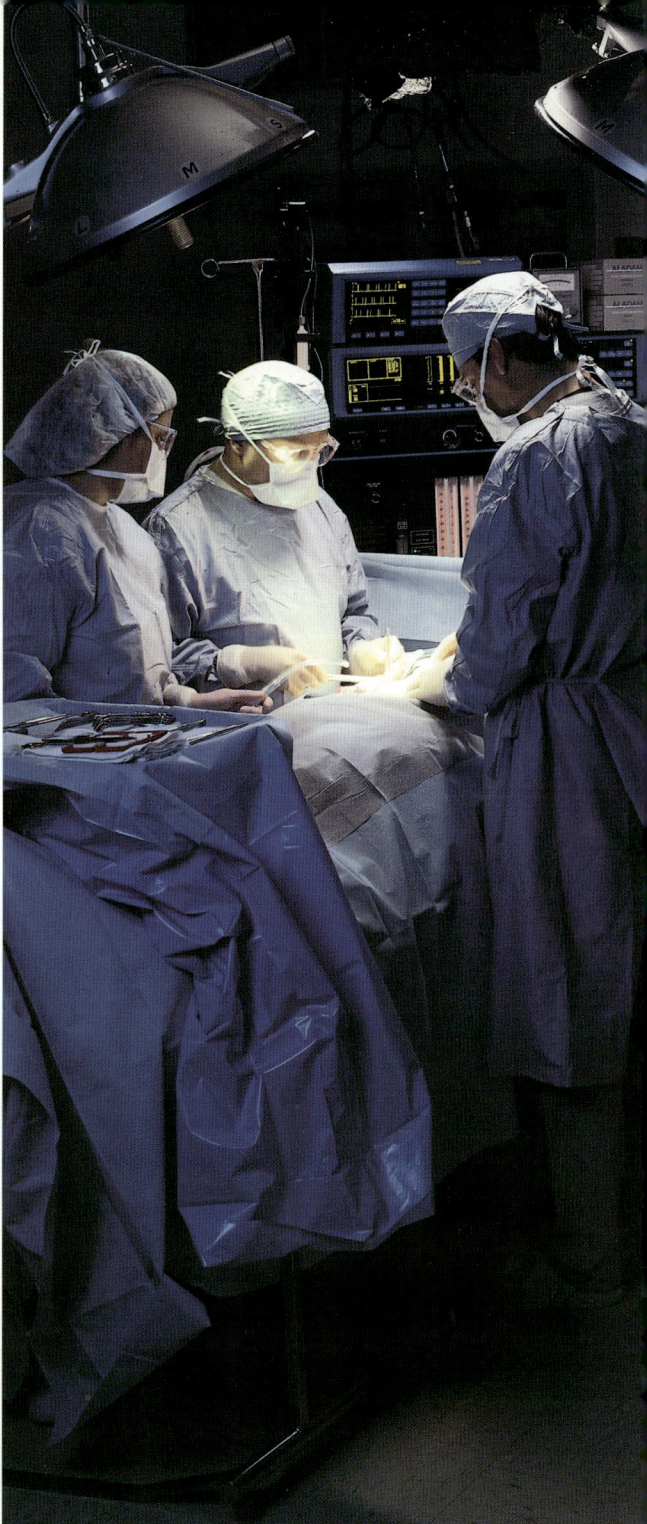

When the electricity has only one path around a circuit we call the circuit a **series circuit**. When it has more than one path we call it a **parallel circuit**. Parallel circuits can deliver more electricity more quickly than a series circuit. This means the bulbs can be brighter. Also, even if there is a break in one part of a parallel circuit the electricity may still be able to flow along another pathway.

1 What is a series circuit?
2 What is a parallel circuit?
3 List the advantages of a parallel circuit for the operating theatre.

Key words

series circuit
parallel circuit

13.4 More power

Switches control whether electricity flows – or not. The wires in the circuit control how much electricity flows. The amount of electricity flowing is called the **current**. We use an **ammeter** to measure the current flowing. The more electricity flowing, the bigger the reading on the ammeter. An ammeter gives a reading in **amps**.

Wires **resist** the flow of current. Insulators resist it so well no electricity gets through at all. Conductors resist it a bit so some of the current is wasted as heat.

- *Why do power station engineers need to know how much electricity flows out from the station?*
- *Do the engineers want power lines with a low or a high resistance? Why?*

Dimmer switches control the resistance in a circuit. This lets us control how brightly a light shines.

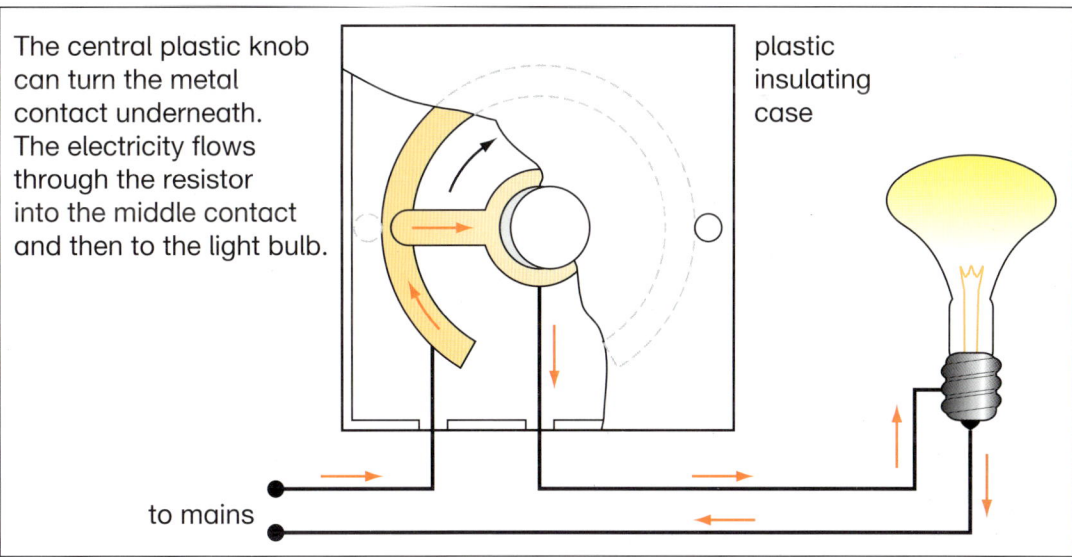

The central plastic knob can turn the metal contact underneath. The electricity flows through the resistor into the middle contact and then to the light bulb.

plastic insulating case

to mains

1 What happens to the brightness of the bulb when the current is bigger?
2 Suggest two ways to make the bulb brighter.
3 What does an ammeter measure?
4 What is the resistance of a wire?
5 What is a dimmer switch?

Key words

current
ammeter
amp
resist
dimmer

14 Power

14.1 Batteries

Batteries come in lots of different shapes and sizes. Some are **rechargeable** and others must be thrown away when they have been used. Different batteries are used for different jobs. These electric cars use giant batteries that weigh more than their engines.

- *What do you use them for?*
- *How do you decide which kind to use?*
- *What is good about using rechargeable batteries?*
- *What is bad about using rechargeable batteries?*

Every battery has two **terminals**. Inside the battery is a liquid or paste between the terminals. This is called the **electrolyte**. **Chemical reactions** inside the battery make electricity flow along a wire connected to the terminals.

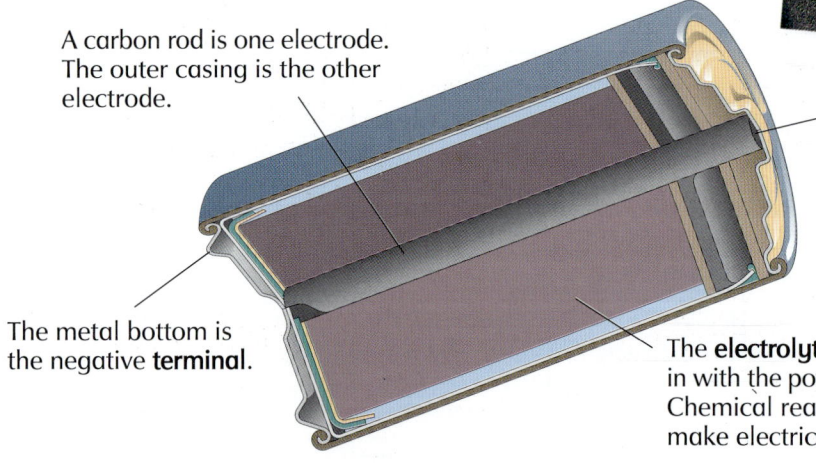

A carbon rod is one electrode. The outer casing is the other electrode.

The metal top is the positive **terminal**.

The metal bottom is the negative **terminal**.

The **electrolyte** is the liquid mixed in with the powder to make a paste. Chemical reactions in this paste make electricity flow.

The **voltage** of a battery is the amount of push it gives to the electricity. Batteries usually have the voltage written on them. You can use a **voltmeter** to check the voltage of a battery.

1 What is a battery?
2 Draw a picture to show how you could use a glass of salty water and pieces of iron and copper to make a battery.
3 Label the two terminals and the electrolyte in your battery.
4 How would you find the voltage of your battery?
5 Draw and label a picture of the inside of a torch battery.

Key words

battery
rechargeable
terminal
electrolyte
chemical reaction
voltage
voltmeter

14.2 What is electricity?

It's easy to put out the fire – just throw water over it! But what happens if the water is 20 metres away from the fire?

- *What would happen if the buckets were only half full?*
- *What happens if the people used bigger buckets?*
- *What would happen if half of the people were missing or dropped the buckets?*

The way water gets to the fire is a bit like the way electricity flows along a wire. The water movement depends on the amount of water in each bucket and how well the people pass the buckets along the chain. The flow of electricity along a wire depends on the **voltage** and the **resistance**.

The voltage is a measure of how much 'kick' the electricity gets when it enters the wire. A large voltage will usually push lots of electricity along a wire.

In the fire chain some of the water will spill. Some buckets may be dropped. The human chain will resist the flow of the water. The amount of water getting to the fire will be less than the amount picked up at the tap!

In the same way, wires resist the flow of electricity. A wire with a big resistance cuts down the amount of electricity flowing. The amount of electricity flowing along a wire is called the **current**. Current is measured in **amps.**

1 What two things affect how much electricity flows along a wire?
2 What does the word resistance mean?
3 Give two ways to increase the flow of electricity to a bulb.

Key words

voltage
resistance
current
amps

14.3 Three-pin plugs

Batteries are fine for small things like radios and torches. But for bigger items you're better off getting electricity from the **mains supply**. Remember: you still need to be careful with electricity!

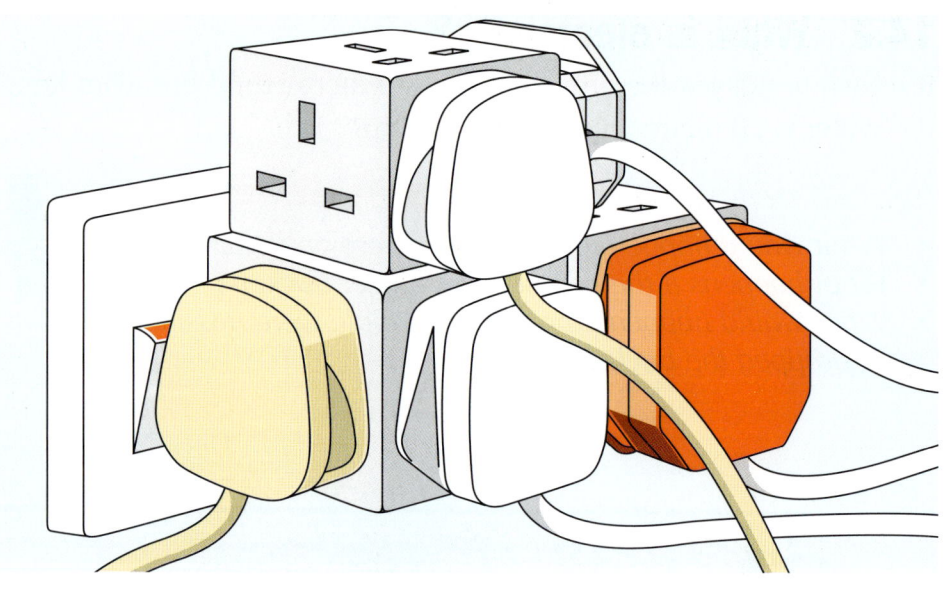

- *How many **sockets** have you got in your bedroom? Is this enough?*
- *How many more sockets do you think you would need?*
- *Which room in your home has the most sockets? Is this enough?*
- *Why is the socket in the picture above dangerous?*

The **plug** in the drawing below is designed so that it is safe to use. The electricity flows along the two wires called **live** and **neutral**. The third wire, called the **earth** wire, is there for safety. Electricity only flows along the earth wire when something has gone wrong with the circuit. Not all plugs have an earth wire – only items with metal parts need an earth.

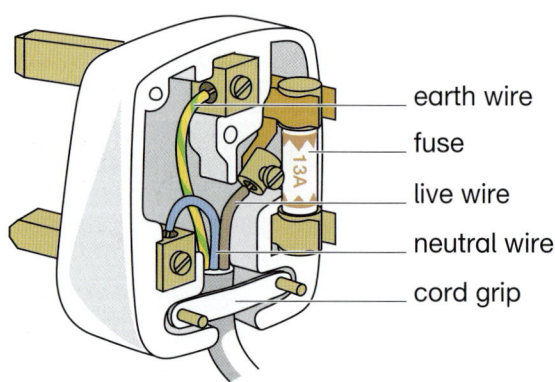

- earth wire
- fuse
- live wire
- neutral wire
- cord grip

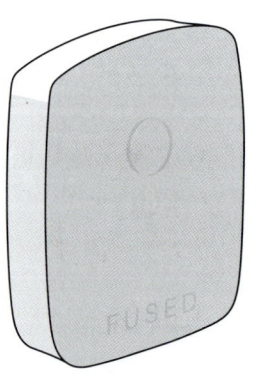

1 Which parts of the plug are metal so that the electricity can flow through them easily?
2 Which parts are plastic so that the electricity can't flow through them?
3 What is the **fuse** for?
4 Why does the plug have a cord grip?
5 Draw and label a diagram of a three-pin plug. Be careful to make the wires the right colours.

14.4 Paying the price

Electricity costs money. The more you use, the more you pay. An **electricity meter** records how much electricity you have used. This lets the electricity company work out how much you owe it. The electricity company sells electricity in **units**. One unit is about enough to run a single bar electric fire for one hour. Things that give out heat tend to use the most electricity.

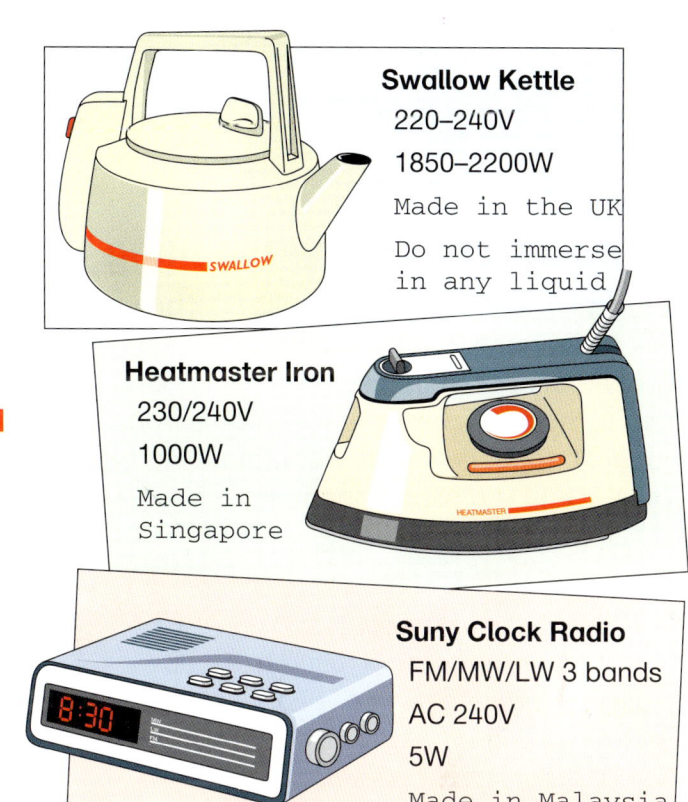

Swallow Kettle
220–240V
1850–2200W
Made in the UK
Do not immerse in any liquid

Heatmaster Iron
230/240V
1000W
Made in Singapore

Suny Clock Radio
FM/MW/LW 3 bands
AC 240V
5W
Made in Malaysia

- *List the things you use at home that use electricity.*
- *Sort them so that the ones that use the most electricity come first.*
- *List some ways to save electricity – and so save money!*

We can work out exactly how much electricity something uses by looking at its **power** rating. Powerful things use a lot of electricity. Less powerful things use less electricity. Power is measured in **watts**.

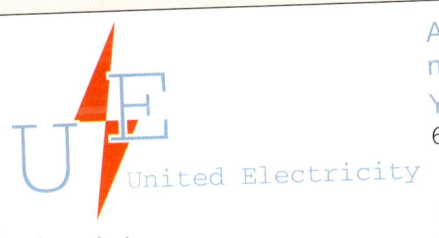
United Electricity

Accounts advice telephone number 01234 567890
Your account number
6545 7391 32

Meter reading date	16 June 1996
Date of issue	17 June 1996

Supply to
Mr and Mrs Gary Johnson
17 Main Street
Westgate
Norwich
NR4 7AJ

METER READINGS				
Present	Previous	Units supplied	Domestic	Amount
60197	59741	456	@ 7.22p	£32.92
Fixed charges				£3.22
AMOUNT NOW DUE				£36.14

1 What is power measured in?
2 How long does a unit of electricity last on a single-bar electric fire?
3 How much have you got to pay?
4 What was the date of the meter reading?
5 What was the meter reading?
6 How many units have you used?
7 How much does each unit cost?

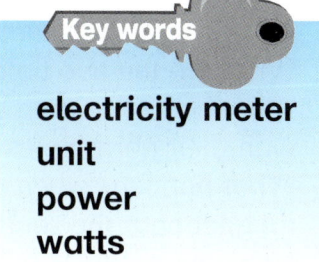

Key words

electricity meter
unit
power
watts

15 Magnets

15.1 Attracting and repelling

Magnets are made of iron or steel. Things which stick to magnets are called magnetic. Only three pure metals are magnetic: iron, nickel and cobalt. Steel is magnetic because it contains iron. The magnetic tape in cassettes contains tiny bits of iron stuck in a plastic ribbon.

- *What sticks to magnets?*
- *How can magnets be used to get bits of metal out of people's eyes?*
- *Have you tried to put two magnets together? What happened?*

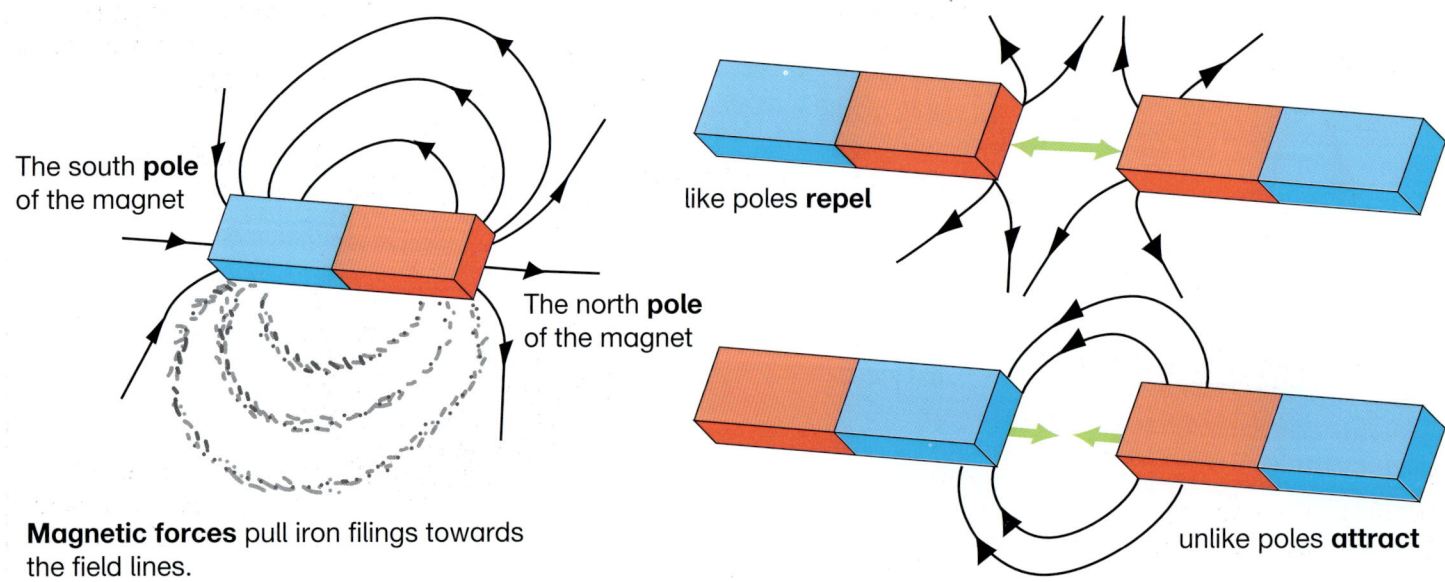

The south **pole** of the magnet

The north **pole** of the magnet

Magnetic forces pull iron filings towards the field lines.

like poles **repel**

unlike poles **attract**

1 List some magnetic materials.
2 What are the two poles of a magnet called?
3 What happens when you put two magnets together with their north poles facing?
4 What happens when you put a north pole of one magnet near the south pole of another?
5 How can magnets help you sort cans for recycling?

Key words

magnet
pole
attract
magnetic force
repel

15.2 Finding your way

Once you are in the trees you depend on a **compass** to move around. You cannot see very far in any direction. You could easily get lost – it's a bit like walking around in a dense green fog.

- *How can travellers find out which way is north?*
- *Have you ever used a map? Where did you go?*
- *Have you ever used a compass? What did you use it for?*

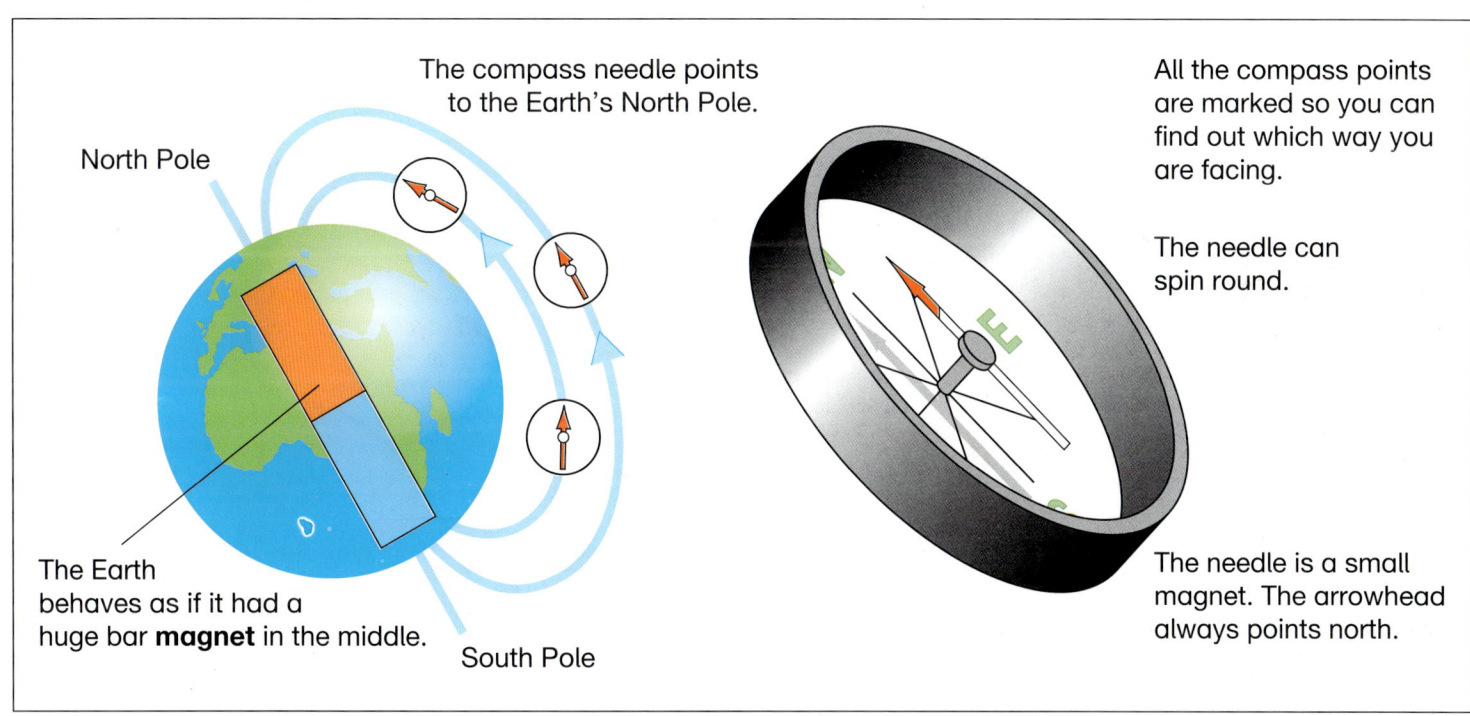

The compass needle points to the Earth's North Pole.

North Pole

The Earth behaves as if it had a huge bar **magnet** in the middle.

South Pole

All the compass points are marked so you can find out which way you are facing.

The needle can spin round.

The needle is a small magnet. The arrowhead always points north.

1 Which of the Earth's poles attracts the north pole of a magnet?
2 Which of the Earth's poles repels the north pole of a magnet?
3 Find out whether you live in the north, south, east or west of your town or village.
4 Which direction is your school from your house?

Key words

compass
magnet

15.3 Attractive scrap

Every year tonnes of metal end up in the scrap yard. Moving them around is easy if you have a crane with an **electromagnet**. You do not need to hook the crane carefully around the bundle of metal – just switch on the magnet and lift.

- *How can you make a magnet strong enough to pick up a car?*
- *Where have you seen an electromagnet at work?*
- *Think of some uses for a magnet you can turn on and off.*

You cannot switch an ordinary magnet on and off. It always pulls with the same **strength**. If you wrap some insulated wire around a piece of metal you can make a magnet you can turn on and off. When electricity flows through the wire the iron bar becomes a strong magnet. When the electricity stops the iron bar is only a weak magnet.

Power pack supplies the electromagnet with electricity. When the electricity is on a **magnetic field** forms. The nail and **coil** behave like a bar magnet.

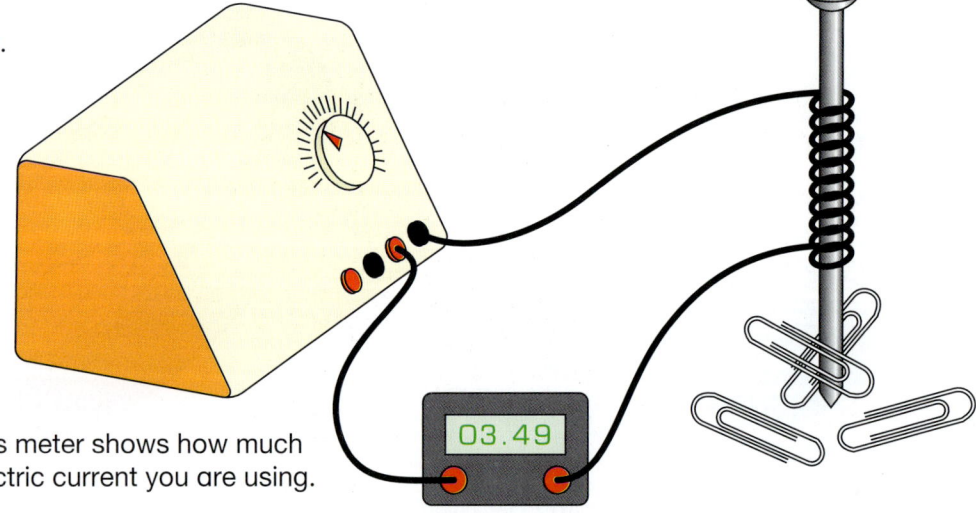

This meter shows how much electric current you are using.

1 What is the main difference between an electromagnet and an ordinary magnet?
2 How do you think electromagnets got their name?
3 List the things electromagnets can be used for.
4 What happens to an electromagnet when you switch off the electricity?
5 List the things you can do to make an electromagnet stronger.

Key words

electromagnet
strength
magnetic field
coil

15.4 Using electromagnets

Glastonbury is the largest open air pop festival in the country. It needs giant speakers to broadcast music to the thousands of fans who turn up. The speakers are in the towers at each side of the stage.

- *Why are the **loudspeakers** held in towers?*
- *How many loudspeakers do you have at home?*
- *Which loudspeaker is the biggest?*
- *Which loudspeaker is the smallest?*

Electromagnets in loudspeakers make a **cone** of paper **vibrate.** The cone produces **sounds** in the air.

The cassette player sends electrical signals to the coil. Signals switch the electromagnet on and off. The large permanent magnet helps to push and pull on the small coil.

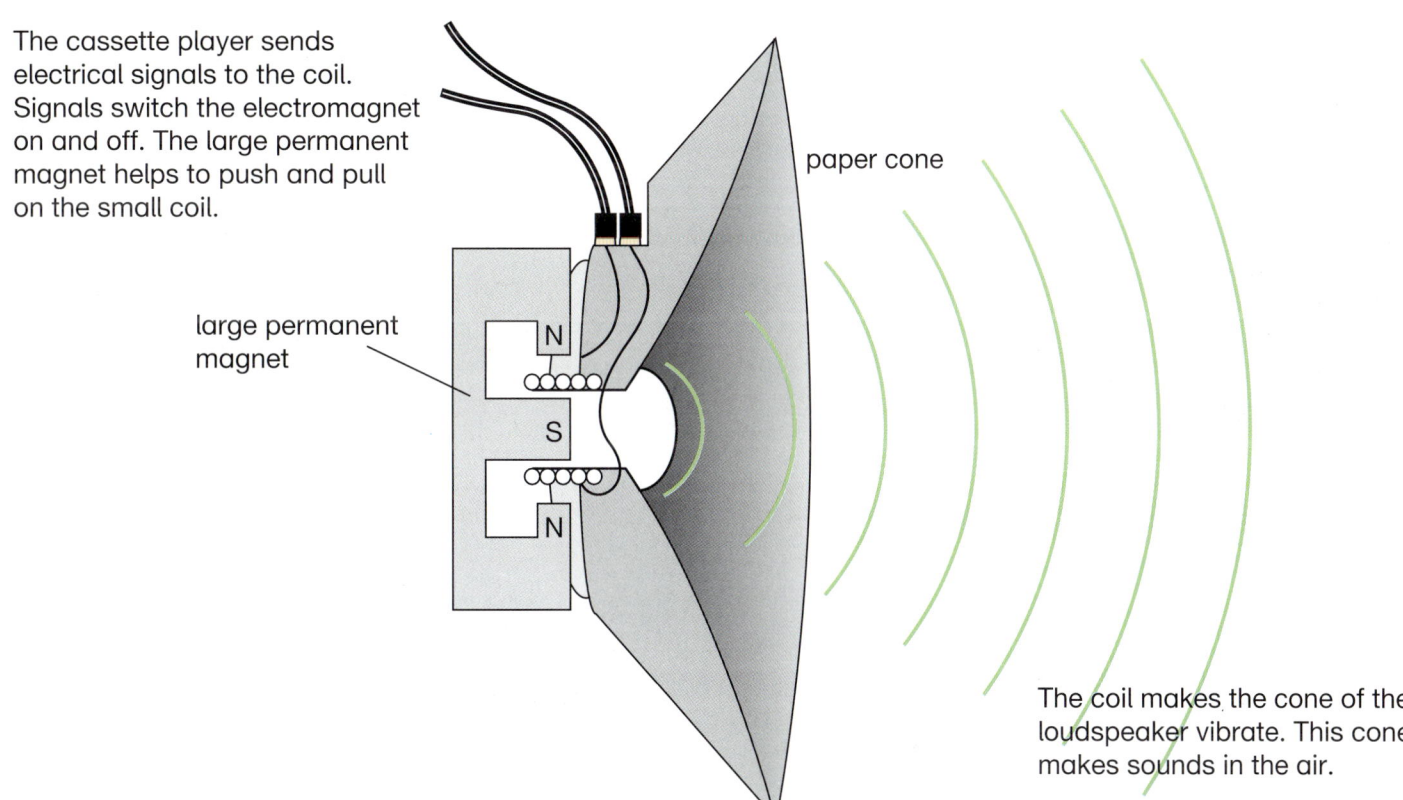

paper cone

large permanent magnet

The coil makes the cone of the loudspeaker vibrate. This cone makes sounds in the air.

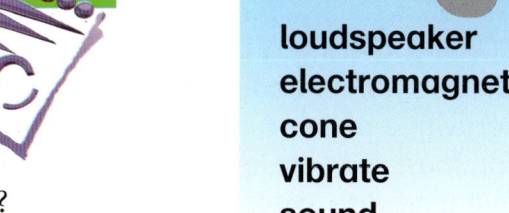

1 List five things which use a loudspeaker.
2 Sort your list into home, school and other uses.
3 How does the electric signal get to the loudspeaker?
4 Which bit of the loudspeaker makes the sound?
5 What happens to the cone when the coil becomes a magnet?
6 What happens to the coil when the signal reaches the speaker?

Key words

loudspeaker
electromagnet
cone
vibrate
sound

16 G-force

16.1 Gravitational landings

You just attach a piece of meat to a hook and dangle it over the end of a speedboat. Wait patiently until you hook a large fish. But make sure you are tied down in your seat or the fish will pull you overboard!

- *Have you ever been fishing?*
- *Where does the pull on the fisherman's line come from?*
- *What will happen to the fisherman if the line suddenly snaps?*

The angler pulls against the fish. We call the pull a **force**. A force can be either a pull or a push. We measure force in units called **newtons**. A force called **gravity** pulls the fish towards the Earth. Gravity pulls on the fish with a force of roughly 10 Newtons for every kilogramme of its weight.

We normally measure weight in kilogrammes. This fish weighs over 60 kg. That gives the angler something to brag about!

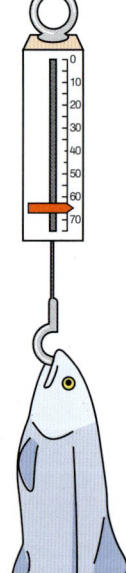

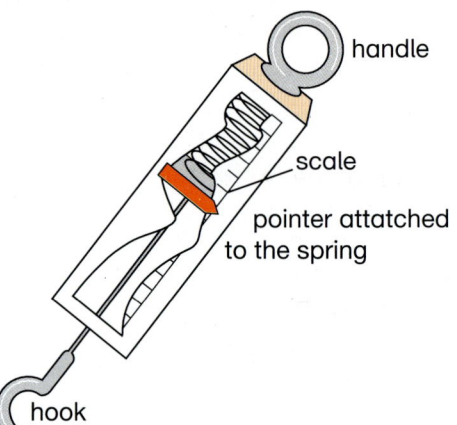

handle

scale

pointer attatched to the spring

hook

1 List some forces you can feel every day.
2 Sort these forces into pushes and pulls.
3 Is gravity a pull or a push force?
4 Why must shop scales be accurate?
5 Build a **newton meter**. Test it with different weights to see how accurate it is.

Key words

force
newtons
gravity
newton meter

16.2 Bouncing back

Some people pay good money to tie themselves to a bridge with a piece of elastic and throw themselves off! They call it bungee jumping. They must be mad!

- *What happens when a bungee jumper falls?*
- *Would you trust a bungee rope?*
- *Would you ever do a bungee jump?*

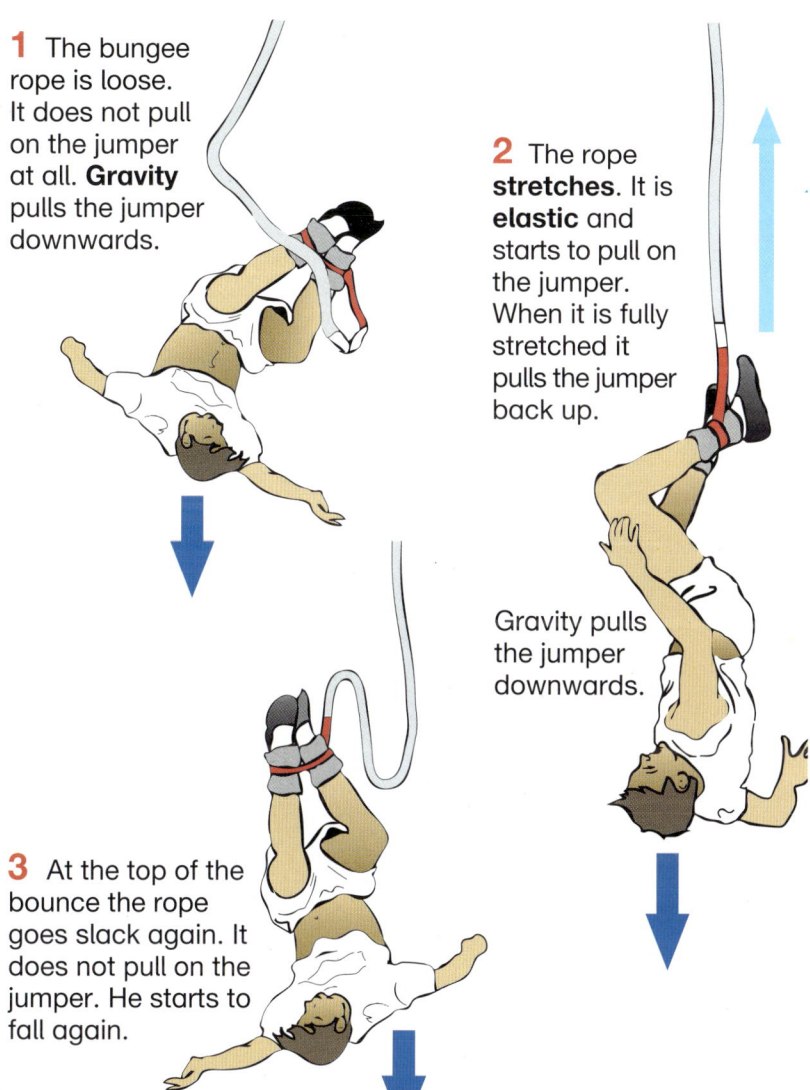

1 The bungee rope is loose. It does not pull on the jumper at all. **Gravity** pulls the jumper downwards.

2 The rope **stretches**. It is **elastic** and starts to pull on the jumper. When it is fully stretched it pulls the jumper back up.

Gravity pulls the jumper downwards.

3 At the top of the bounce the rope goes slack again. It does not pull on the jumper. He starts to fall again.

1 What pulls the bungee jumper towards the ground?
2 What stops the jumper hitting the ground?
3 Bungee ropes are elastic. What does elastic mean?
4 Is a bungee rope that is twice as thick as another twice as strong? Test out your answer with elastic bands and weights.

Key words

gravity
stretch
elastic

16.3 Coming down to Earth

Parachutes let people and things fall to the ground slowly so they are not hurt. Sometimes parachutes are used to drop food and machines to isolated areas.

* *Would you ever go sky diving?*
* *You always have to pack your own parachute before you jump. Why do you think sky divers have this rule?*

The parachute collects air. The air pushes up. This is called **air resistance**. Air resistance balances some of the pull of **gravity**. The parachute slows down the person's fall.

Gravity pulls the parachutist downwards.

1 List some sports which use gravity to pull things down.
2 Sort the sports into ones you have tried and ones you have not.
3 Why do hang gliders use very large wings?
4 Why do birds like vultures have big wings?

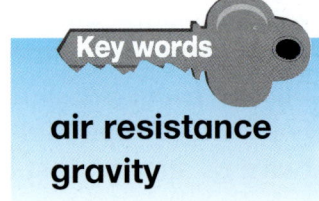

Key words

air resistance
gravity

16.4 Playing safe

Every year children have accidents in playgrounds. Modern adventure playgrounds often have climbing frames and skateboard runs. This makes them more fun – and more dangerous! The well-equipped young skateboarder now has a helmet, knee pads and elbow pads.

- *How might children hurt themselves in an adventure playground?*
- *Have you ever been on an assault course?*
- *Why don't divers get hurt when they land in water?*

Gravity pulls you down when you fall. The harder the ground the more likely you are to hurt yourself. Soft surfaces are less painful.

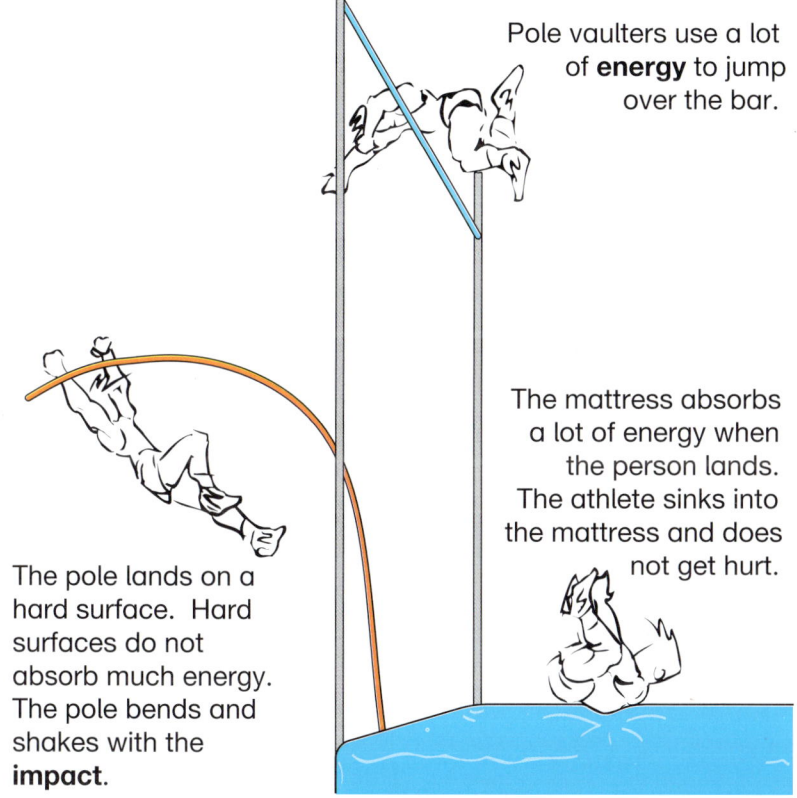

Pole vaulters use a lot of **energy** to jump over the bar.

The mattress absorbs a lot of energy when the person lands. The athlete sinks into the mattress and does not get hurt.

The pole lands on a hard surface. Hard surfaces do not absorb much energy. The pole bends and shakes with the **impact**.

1 Find out what kind of surfaces your local playgrounds have.
2 What kind of injuries do children get in adventure playgrounds?
3 Do you think the kind of surface can cut down the number of serious accidents in the playground?
4 How else can children be protected while they play?

Key words

energy
impact

17 Theatres

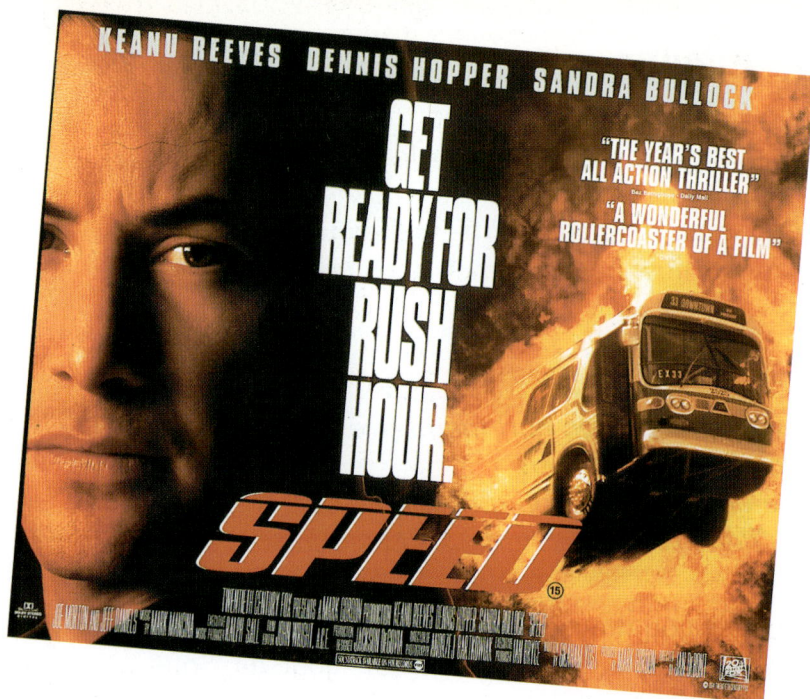

17.1 Eyes

In 1995 the film *Speed* took millions of dollars worldwide. Over ten million people went to see it and one person went more than fifty times! All this depends on patterns of light bouncing off a flat, white screen.

- *What was the last film you went to see?*
- *Was it any good? Why? Why not?*
- *Would you go to see it again?*

We see things when light goes into our eyes. Some things make their own light. We call these things **luminous** objects. So, a light bulb and a candle flame are luminous objects. **Non-luminous** objects cannot make their own light. They reflect the light made by something else. The screen in a cinema is non-luminous. It reflects coloured light from the projector.

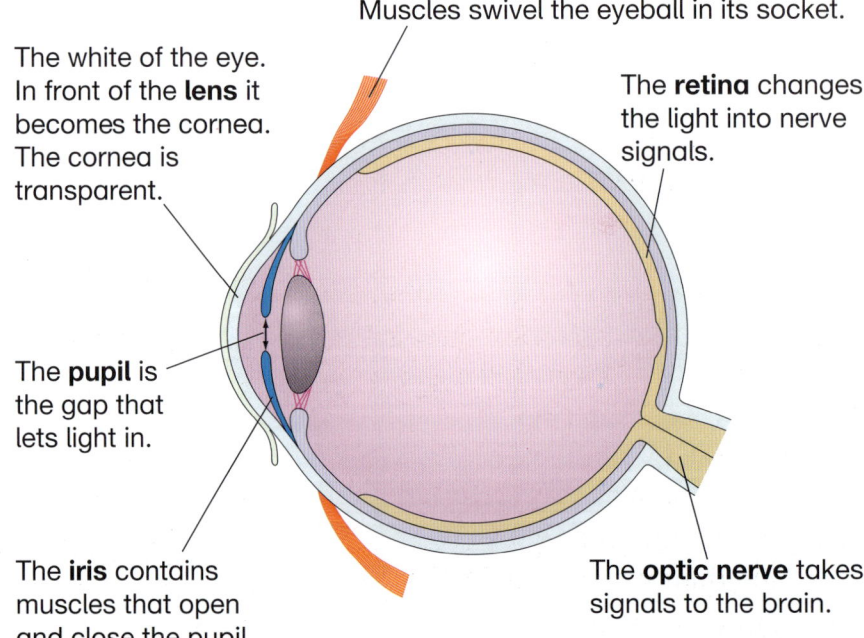

Muscles swivel the eyeball in its socket.

The white of the eye. In front of the **lens** it becomes the cornea. The cornea is transparent.

The **retina** changes the light into nerve signals.

The **pupil** is the gap that lets light in.

The **iris** contains muscles that open and close the pupil.

The **optic nerve** takes signals to the brain.

1. What does luminous mean?
2. Name three things that are luminous.
3. How do we see non-luminous objects?
4. Name three things that are non-luminous.
5. List the parts of the eye and explain what they do.

Key words

luminous
non-luminous
lens
pupil
cornea
retina
optic nerve
iris

17.2 Reflections

Mirrors are special reflectors. They **reflect** light in particular directions. A movie screen reflects light in all directions. This paper reflects light in all directions. The mirrors in the photograph reflect the Sun's light onto one small place – and it gets hot!

- *How many mirrors have you got at home?*
- *What else can give you a clear reflection?*

We can use **rays** of light to investigate how mirrors reflect. A ray of light is a thin **beam**. We can see this beam when it touches a piece of white paper.

Mirrors are more complicated than they look. Mirror writing is writing that is the right way up but back to front – it makes no sense. But look at it in a mirror. This is because mirrors flip images – so the right side turns up on the left and the left on the right.

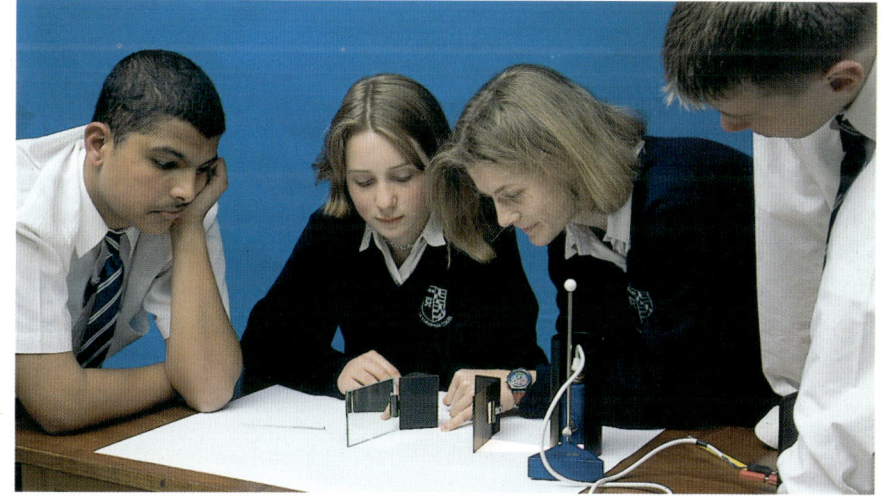

1   What is the difference between the way a mirror and a piece of white paper reflect light?
2   Use a light beam to find out how mirrors reflect light. Try to find a rule that tells you what direction a beam will bounce off a mirror.
3   Write your name in mirror writing.
4   Draw a diagram to show how a flat mirror reflects a ray of light.

Key words

reflect
ray
beam

17.3 Light shows

When light can't get through something it makes a **shadow**. In Indonesia shadow puppets are used to tell traditional stories. This is probably the oldest cinema in the world!

- *How could you make the puppets look as if they were changing size on the screen?*
- *How is a shadow puppet show different from a black and white movie film?*

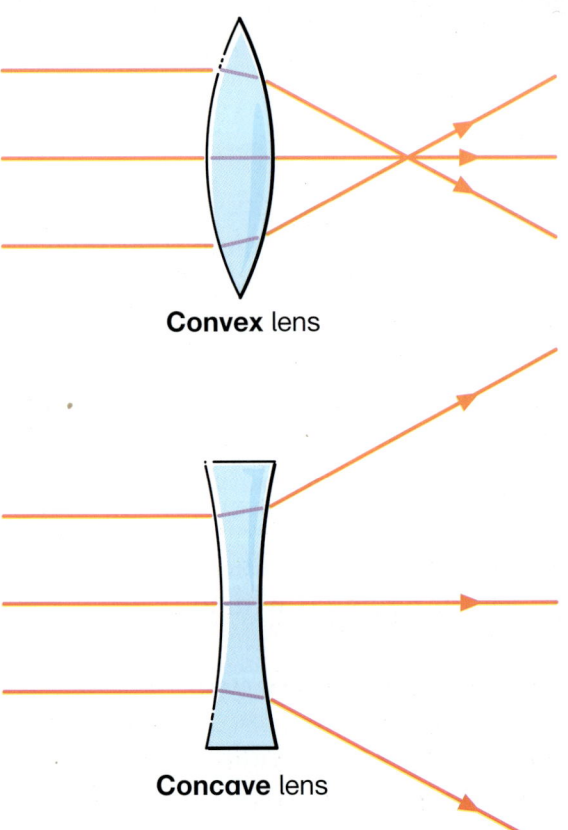

Convex lens

Concave lens

Light travels in straight lines. Mirrors reflect light. **Lenses** change its direction. So lenses bend light. The lens in the eye bends the rays of light to make an image on the retina. In a camera the lens bends light to make an image on the film. In a film projector the lens bends the rays of light to make an image on the screen.

1 How many people do you think were needed to make the shadows in the photograph?
2 Make your own shadow puppet and test it.
3 How can you make the shadow bigger?
4 How can you make it smaller?
5 Use a light box and different lenses to find out how much they change the light beam's direction.

17.4 Coloured lights

Sunlight is called white light. It contains all the other colours mixed in together. A triangular piece of glass called a **prism** can sort these out into a **spectrum**. The prism lets all the colours through but spreads them out.

Coloured spotlights shine ordinary light through a piece of coloured plastic. The plastic only lets one colour through and **absorbs** the rest.

- *Where have you seen a spectrum?*
- *What caused the spectrum?*
- *Street lights often have an orange glow. What does this make white cars look like at night?*

The red candle in the photograph looks red because it **reflects** red light. It absorbs all the other colours. The white candle looks white because it reflects all the colours of light.

Here is the same scene in red light. The white candle reflects the red light and so looks red. The blue candle absorbs the red light. There is no blue light to reflect so it looks black.

1 Draw a picture to show how you would use a torch and a piece of green plastic to make green light.
2 Which candles look red in photo 2? Why?
3 Which candle has the darkest colour in photo 2? Why?
4 Draw a picture of a green apple on a white plate seen in normal white light.
5 Now draw the same apple and plate seen in green light.

Key words

prism
spectrum
absorb
reflect

18 The final frontier

18.1 Rocket science

The most exciting journey of all time began in Cape Kennedy, Florida on July 16th 1969. Three men started on their journey to the Moon.

- *Why go to the Moon? Can you think of a reason?*
- *The space project cost billions of dollars. Some people say it was a waste of money. Do you agree? Why?*

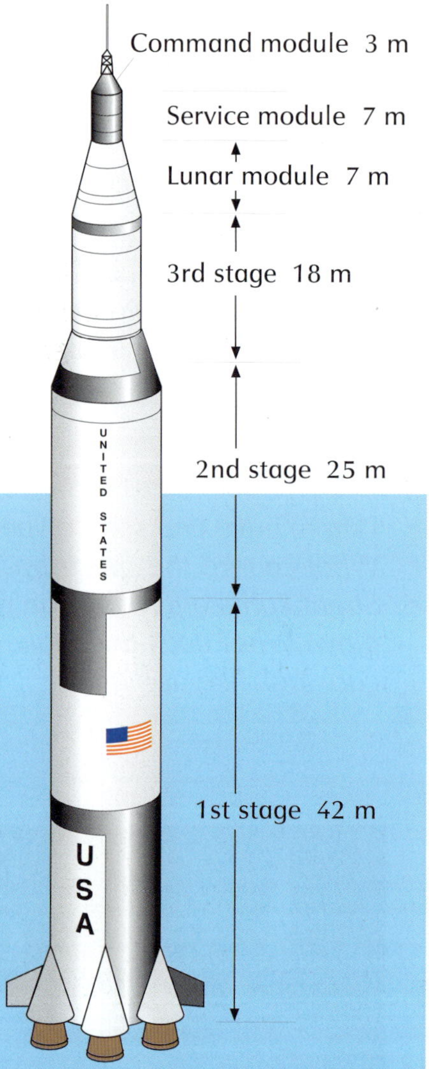

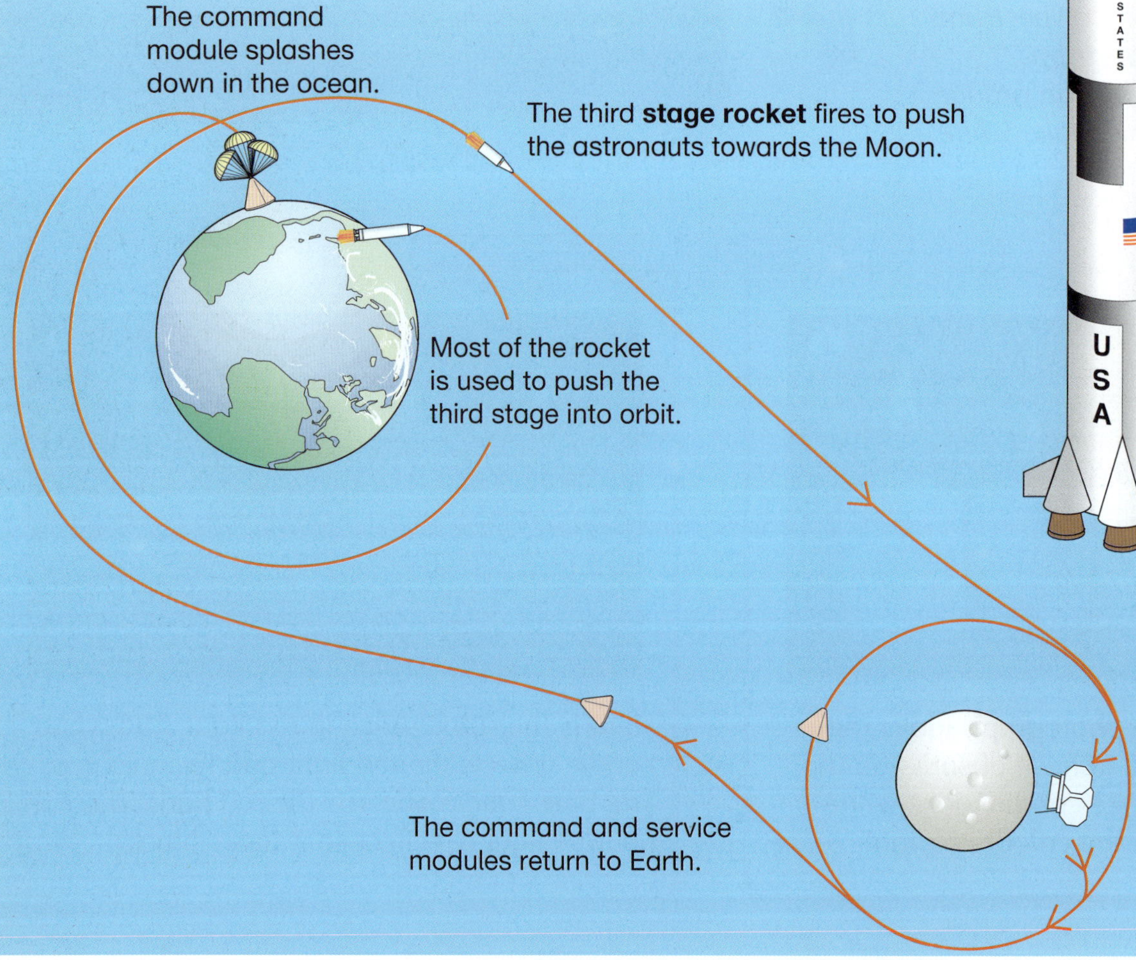

Command module 3 m
Service module 7 m
Lunar module 7 m
3rd stage 18 m
2nd stage 25 m
1st stage 42 m

The command module splashes down in the ocean.

The third **stage rocket** fires to push the astronauts towards the Moon.

Most of the rocket is used to push the third stage into orbit.

The **lunar module** lands on the Moon.

The command and service modules return to Earth.

1 Which part of the rocket landed on the moon?
2 Which part of the rocket returned to Earth?
3 How tall is the lunar module?
4 Why is the rocket so large compared with the lunar module?

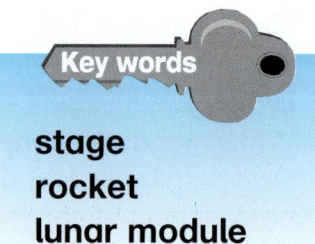

Key words

stage
rocket
lunar module

18.2 Moon landing

After five days' flight Apollo 11 reached the Moon. At 2.56 on the morning of July 21st 1969 Neil Armstrong stepped onto the surface. His first words were heard in every part of the Earth. 'That's one small step for a man, a giant leap for mankind.'

- *One day there may be package holidays to the Moon. Would you like to go? Why? Why not?*
- *What do you think Neil Armstrong felt like as he climbed down the ladder to the Moon's surface?*

The Apollo missions

Mission	Year	Notes
11	1969	First men on the Moon
12	1969	15 hours on the Moon
13	1970	Spacecraft struck by a **meteorite**. Landing cancelled.
14	1971	44 kg of Moon rock brought back.
15	1971	Lunar Rover is used – the first vehicle for driving on the Moon's surface.
16	1972	
17	1972	The last time men went to the Moon.

The Moon's surface is covered with **craters**. These are made when meteorites crash into the Moon. The Moon has no atmosphere and so the meteorites do not burn up as they fall. Meteorites may be the remains of an exploded star.

The Moon's gravity pulls the meteorite towards the surface.

When the meteorite hits the surface it produces a crater.

There is no wind or rain on the Moon so the crater is not worn away.

1 There is no wind or rain on the Moon. What would happen to the footprints left by an astronaut?
2 How many trips have there been to the moon since Apollo 11?
3 Which trip nearly ended in disaster?
4 Plan an investigation to find out how the size of a meteorite affects the size of a crater.

Key words

meteorite
crater

73

18.3 The space shuttle

The shuttle is the first space ship to be used more than once. It uses a rocket motor to go into **orbit** and then glides back to Earth like a plane.

The shuttle needs a push to move it into orbit. The **booster rockets** supply the push. The shuttle is going forwards and upwards.

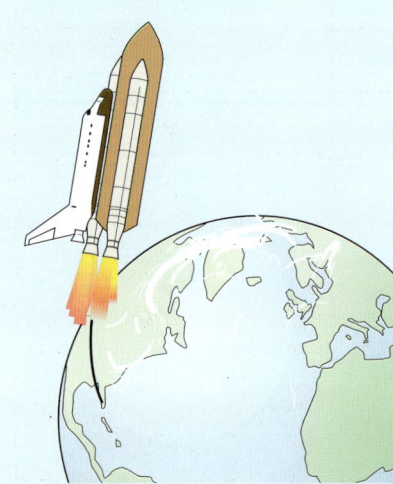

Once the shuttle reaches a certain height the rocket motors are switched off. The shuttle goes forwards but does not go any higher. The shuttle is falling around the Earth.

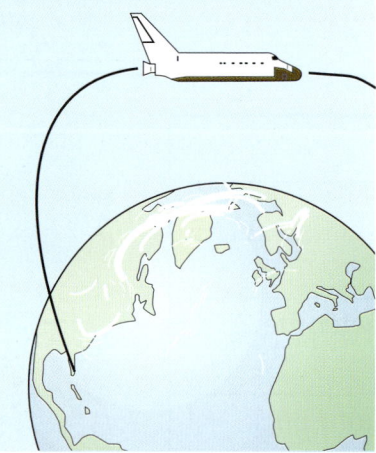

All the objects in the shuttle, including the astronauts, are falling at the same speed. Everything that is not fixed down floats around in the shuttle. Astronauts call this **weightless** conditions.

1 Why do the astronauts float around inside the space shuttle when it is in orbit around the Earth?
2 What kind of experiments might be carried out on the space shuttle when it is in orbit?
3 What problems do you think astronauts have on a long flight?

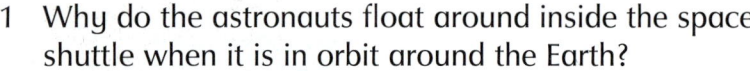
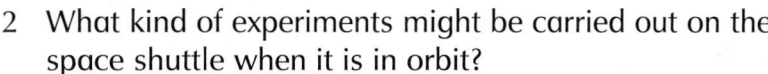

Key words

orbit
booster rocket
weightless

74

18.4 The Solar System

The last Moon shot was in 1972. The next step, sending astronauts to Mars and bringing them back safely, is much more difficult. The distance to Mars is much greater than to the Moon. And Mars is the nearest planet to Earth!

A view of Mars.

Planet	Distance from Sun /million km	Diameter /km	Weight /Earth=1	Average surface temperature /°C	Time to travel round Sun /years
Mercury	58	4969	0.05	510	0.24
Venus	108	12200	0.8	480	0.62
Earth	150	12757	1.0	15	1.00
Mars	228	6800	0.1	-50	1.88
Jupiter	779	143600	318	-250	11.86
Saturn	1427	121000	95	-180	29.46
Uranus	2670	47000	15	-220	84.0
Neptune	4496	44600	17	-200	164.8
Pluto	5906	3000	0.06	240	247.7

Views of Venus.

1 What is at the centre of the **Solar System**?
2 How many **planets** are there in the Solar System?
3 Which planet is closest to the Sun?
4 Which is the largest planet?
5 Why is it so much more difficult to send astronauts to Mars than to the Moon?

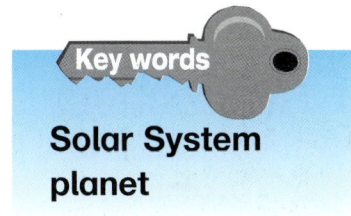

Glossary

abrasive	rough
absorb	to take in
acid	a chemical which turns litmus paper red
adaptation	something about an animal that makes it suited to its environment
afterbirth	the placenta and membranes that leave the mother's womb after the baby has been born
air resistance	air pushing against an object
alkali	a chemical which turns litmus paper blue
alloy	a mixture of two or more metals
aluminium	a light, silver-coloured metal
ammeter	meter used to measure electric current
amp	unit used to measure electric current
antenatal	before birth
antifreeze	liquid containing a chemical that makes its freezing point very low
Apatosaurus	large extinct plant-eating dinosaur
artery	blood vessel carrying blood away from the heart
atom	smallest part of an element
atria	two top chambers of the heart
attract	pull together
bacteria	microscopic living things
battery	device which stores chemical energy and gives out electricity
bladder	body organ which stores urine
blood pressure	the force of blood in the blood vessels
blood transfusion	injecting blood into someone's body
blood vessel	tube that carries blood in the body
boiling point	temperature at which something boils
capillary	thin-walled blood vessels that carry blood from arteries to veins
carbohydrates	chemicals which give us energy
cell	the basic unit of life
cell membrane	very thin outer covering of a cell
cervix	muscular entrance to the womb
chemical reaction	when two or more chemicals react together, bringing about a change
circuit	a loop of electrical conductors
circulation	transport of blood or other substances around the body
coil	wire made into a springy shape
colour fast	colour will not run or leak out when wet
coma	deeply unconscious
compound	substance made of two or more elements joined chemically
concave	a concave lens is thinnest in the middle
condense	to change from a gas or vapour to a liquid
conductor	a material that lets electricity go through it
contraction	rhythmic shortening of the muscles of the womb
convex	a convex lens is fattest in the middle

cornea	clear front of the eye
corrode	to rust
current	flow of electricity
cytoplasm	see-through jelly-like substance in cells
detergent	cleaning agent
diabetes	a disease which makes it hard to control your blood sugar level
diarrhoea	loose bowels
diet	the food and drink you eat
digestion	the process of breaking food down in the gut
dilate	to make wider
dinosaur	extinct giant reptile
displacement reaction	when one metal replaces another in a solution
dissolve	sugar dissolves in water when it disappears into the water
donor	the person who gives the organ in a transplant
electric current	flow of electricity
electrolyte	liquid or paste inside a battery
electromagnet	a magnet which only works when electricity flows through it
electroplating	covering something with metal by using electricity
element	substance containing only one type of atom
energy	the ability to do something
enzymes	substances which change the rate of chemical reactions
evaporate	change from liquid to gas or vapour
excretion	to get rid of waste from the body
expand	get bigger
extinct	a type of plant or animal which has died out
fat	chemicals that give energy and warmth
fertilises	joining of one sperm and one egg
fever	a high temperature in the body
fibre	1 food that cannot be digested
	2 thread which can be woven
flameproof	fabric treated so it will not easily catch fire
force	push or pull
forelimb	front leg of an animal (or arm)
fossil	preserved remains of an animal or plant
freezing point	the temperature at which a liquid freezes
friction	rubbing force which produces heat and wear and tear
fuse	safety device to stop electricity flowing
gene	contains the information for particular characteristics
global warming	the Earth's atmosphere becoming hotter
gravity	a force which pulls objects together
hard water	water that is difficult to make bubbles with
heart attack	when the muscles of the heart stop beating
host	the person who receives the organ in a transplant
hypothermia	when your body gets too cold to work properly
Ichthyosaurus	an extinct fish-like dinosaur
impact	the force with which one thing hits another
indicator	substance showing whether something is acid or alkaline

ingredient	items in a mixture
insoluble	a substance that will not dissolve
insulator	a material that does not let electricity go through it
insulin	a chemical which controls the amount of sugar in the blood
iris	coloured ring of muscle at the front of the eye
kidney	body organ which cleans waste from the blood
lens	1 curved piece of glass or plastic that bends light rays
	2 part of the eye that bends the light rays to make an image on the retina
life process	functions that go on in the living body
liquid	not solid or gas
litmus paper	paper used to detect acids or alkalis
lubricate	to make slippery
luminous	giving off light
lungs	organs in the chest which exchange gases with air
magnet	an object which attracts iron
magnetic field	the force field around a magnet
magnetic force	the force a magnet makes
melting point	the temperature at which a solid melts
minerals	substances got out of the Earth by miners
mixture	substance formed when ingredients are combined
mordant	substance which fixes a dye in a material
neutral	1 not acid or alkaline
	2 blue wire in a three-pin plug
neutralisation	when an acid and an alkali react together to make water
newton meter	an instrument for measuring pulling forces
newton	a unit used to measure force
non-abrasive	smooth
non-luminous	something which reflects light but does not make its own
nucleus	structure in a cell that acts as its headquarters
nutrition	food
oil	fuel obtained from the ground as thick black liquid
optic nerve	carries messages to the brain
oviduct	tube between the ovary and the womb that eggs move through
parallel circuit	an electric circuit that has more than one loop
particle	a very small piece
pH	a measurement of how acidic or alkaline something is
placenta	structure within the womb which passes food and oxygen to the unborn baby
plasma	the liquid part of blood
platelets	small particles in the blood which help it to clot
pole	the ends of a magnet
power	how much energy arrives at a point every second
prism	a piece of glass or plastic that splits rays of light to make a spectrum
properties	a characteristic that belongs to something
proteins	chemical needed by the body for growth and repair
pterodactyl	extinct flying dinosaur

pupil	the part of the eye that lets the light in
ray	thin line of light
reactivity	how easily a chemical reacts with another
rechargeable	a battery which can have electricity put back into it and so can be used again
red blood cells	disc-shaped oxygen-carrying cells in the blood
reflect	to bounce light back
reflex	an automatic response
repel	a force which pushes things apart
reproduce	make more of your species, such as humans making babies
resist	to try to stop
respiration	a chemical reaction which releases energy from food
retina	the part of the eye which detects light
series circuit	an electrical circuit with no junctions
soft water	water that foams easily
solid	not a gas or liquid
soluble	will dissolve
solute	the solid that dissolves in a liquid
solution	a liquid with a substance dissolved in it
solvent	a liquid capable of dissolving a substance
spectrum	coloured light as seen in a rainbow
sperm	a male sex cell
stimulus	a change in the things around you which makes you respond
strength	to be strong
stress	a push or a pull that makes something longer
stretch	to make something longer
substance	what things are made from
sweat	a watery liquid produced by the skin when it is hot
temperature	a measure of how hot something is
terminal	1 one of the ends of a battery
	2 one of the metal prongs of a three-pin plug
transplant	to put a living organ into another person
Tyrannosaurus	an extinct meat-eating dinosaur
umbilical cord	tube containing blood vessels that joins the baby to the placenta
ureter	a tube which carries urine from the kidneys to the bladder
urine	a pale yellow liquid made by the kidneys from water and waste in the blood
vagina	muscular tube that links the womb to the outside of the woman's body
valve	a flap that stops the blood from flowing backwards
vein	blood vessel that carries blood towards the heart
ventricle	lower chambers of the heart that pump blood into the arteries
vibrate	move from side to side while staying in one place
viscosity	thickness of liquid
vitamins	substances needed by the body in small amounts
voltage	the amount of push that a battery gives to the electricty going through it
voltmeter	a device to measure voltage

water vapour	water in the state of a gas
water-resistant	lets only a little water through
waterproof	lets no water through
watts	unit of measurement of the power of an electric device
weight	how heavy something is
white blood cells	colourless cells in the blood that protect the body from disease
windproof	lets no wind through
womb	female organ where the baby develops
yarn	threads twisted together